Lecture Notes in Mathematics 1502

Editors:
A. Dold, Heidelberg
B. Eckmann, Zürich
F. Takens, Groningen

Carlos Simpson

Asymptotic Behavior of Monodromy

Singularly Perturbed Differential Equations on a Riemann Surface

Springer-Verlag
Berlin Heidelberg New York
London Paris Tokyo
Hong Kong Barcelona
Budapest

Author

Carlos Simpson
Laboratoire de Topologie et Géométrie, URA
Université Paul Sabatier, U. F. R. M. I. G.
31062 Toulouse-Cedex, France

Mathematics Subject Classification (1991): 34D15, 14E20, 30E15, 34B25, 41A60

ISBN 3-540-55009-7 Springer-Verlag Berlin Heidelberg New York
ISBN 0-387-55009-7 Springer-Verlag New York Berlin Heidelberg

Printed in Germany

Typesetting: Camera ready by author
Printing and binding: Druckhaus Beltz, Hemsbach/Bergstr.
46/3140-543210 - Printed on acid-free paper

CONTENTS

Introduction

We will study the question of how the solution behaves when an algebraic system of linear ordinary differential equations varies. We will consider a family of ordinary differential equations indexed linearly by a parameter t, different from the parameter of differentiation z. Fix initial conditions at a point $z = P$; then each equation uniquely determines its solution at all other points. The solutions depend on the equations, so they depend on the parameter t. Evaluating the solutions at a point $z = Q$, we get a function $m(Q,t)$. We will investigate the behavior of this family of solutions, as $t \to \infty$ in such a way that some coefficients of the equations go to infinity.

The simplest example of this situation is the family of equations

$$\frac{dm}{dz} - tm = 0.$$

With initial conditions $m(0,t) = 1$, the solution is $m(z,t) = e^{tz}$. For a fixed $z = Q$, this behaves exponentially in t.

We will look at some families of equations generalizing this basic example, allowing $r \times r$ matrix-valued functions for the coefficients and solutions, and considering equations which are holomorphic and defined globally on a compact Riemann surface. The behaviour of the family of solutions will be more complicated, but still approximately exponential in t. It is relatively easy to give a crude exponential upper bound, but more difficult to bound the growth rate from below.

Our goal is to prove the existence of a nonzero asymptotic expansion for $m(Q,t)$, a sum of terms of the form exponential in t times asymptotic series in negative fractional powers of t and powers of $\log t$. An asymptotic expansion gives a precise estimate on the growth of $m(Q,t)$, showing for example that the growth is exponential rather than sub-exponential. The families we consider will be subject to some assumptions which simplify but do not trivialize the problem. Thus, our theorem is not the most general one which could be hoped for, but it is close to the general case and should serve to illustrate what happens.

I

We consider holomorphic systems of first order differential equations on a compact Riemann surface S, and assume that the underlying vector bundles are trivial, so the systems of equations are given by matrices of one-forms. We will look at an

[1]Supported by NSF grant DMS-8705757.

affine linear family of matrices of one-forms indexed by parameter t. An equation in the corresponding family can be be written as

$$dm = (tA + B)m$$

where m is an $r \times r$ matrix of unknown holomorphic functions, $A = (a_{ij})$ and $B = (b_{ij})$ are $r \times r$ matrices of holomorphic one-forms on S, and the parameter t is a complex number. Our simplifying assumptions on the family are that A is a diagonal matrix with distinct entries, and that B has zeros along the diagonal. If z is a local coordinate on the Riemann surface, and if the entries in the matrices A and B are given by $a_{ij} = \alpha_{ij}(z)dz$ and $b_{ij} = \beta_{ij}(z)dz$ respectively, then the system of equations can be written in the more usual form

$$\frac{\partial m_{ik}(z)}{\partial z} = \sum_j (t\alpha_{ij}(z) + \beta_{ij}(z))m_{jk}(z).$$

Fix a point P and consider the initial conditions $m_{ij}(P) = \delta_{ij}$. Then there is a unique solution with these initial conditions, holomorphic but multivalued on S. If Q is another point and if we fix a path from P to Q, then the continuation of the solution along the path is unique. The value of the solution at Q depends analytically on the complex number t, so we obtain an entire holomorphic function $m(t) = m(Q,t)$. The goal is to describe the behavior of this function as t goes to infinity.

Our result is that as t goes to infinity in some fixed direction, say the positive real direction, there is an asymptotic expansion for $m(Q,t)$, meaning an expression of the form

$$m(Q,t) \sim \sum_{i=1}^{r} e^{\lambda_i t} \sum_{j=J}^{\infty} \sum_{k=0}^{K(j)} c_{ijk} t^{-\frac{j}{N}} (\log t)^k.$$

Call the term with $e^{\lambda_i t}$ the term with *exponent* λ_i. In the above expression, the real parts of the exponents $\Re\lambda_i$ are all equal. The statement that a function has an asymptotic expansion does not mean that the series converges to the function, but that any finite partial sum of the series provides an appropriately good approximation to the function for large values of t. In particular if the expansion is identically zero, the statement only amounts to a bound on the rate of growth of the function. Thus a nonzero expansion is to be hoped for. Our expansion will be nonzero for generic choices of the matrix B.

II

In order to illuminate the topological and geometric significance of the asymptotic expansion, we will discuss in §11 a specific example, the example of connections with structure group $S\ell(2)$. In our notation above, this means that A and B are 2×2 matrices with trace zero. In particular, A is determined by a single one-form a. This one-form determines a singular metric $ds = |\Re a|$. For generic values of B, the real parts of the exponents in the asymptotic expansion are equal to the geodesic distance with respect to this metric, from P to Q along the given homotopy class of paths (the length of a path in this metric is just the total variation of the real part $\Re \int_P^x a$).

For general $G\ell(n)$, given any path γ from P to Q, there is a number $\xi_{path}(\gamma)$ which represents the best bound on the exponent in the expansion which can be obtained simply from integrating the differential equation along the path. Roughly, this can be thought of as follows. For nearby points Q_1 and Q_2, the function $m(Q_1, Q_2, t)$ giving the solution at Q_2 with initial conditions equal to the identity matrix at Q_1, has an asymptotic expansion whose exponent is given by the size of the coefficients of the equation between Q_1 and Q_2 (this case of our result has been known for some time [19] [2]). If we choose closely spaced points $P = Q_0, Q_1, \ldots, Q_k = Q$ along the path, then the solution is a product $m(P, Q, t) = m(Q_{k-1}, Q_k, t) \cdots m(Q_0, Q_1, t)$. Formally, the product of the asymptotic expansions gives an asymptotic expansion for $m(P, Q, t)$ whose exponent is the sum of the exponents for each of the terms. This sum of the exponents for small segments along the path is, in the limit as the division is made small, $\xi_{path}(\gamma)$. For long paths the product of the asymptotic expansions will usually be zero, reflecting some cancellation along the way. The true nonzero expansion will occur at lower exponent.

In the $S\ell(2)$ case, $\xi_{path}(\gamma)$ is the total variation as described above. Even in this case, if γ is not geodesic with respect to the total variation metric, then there will be cancellation in the continuation of the solution along the path. This can be seen geometrically in the possibility of analytically continuing the solution while moving the path to one with smaller total variation. One might ask whether this is the general picture, that is whether $\xi_{path} \stackrel{def}{=} \inf_\gamma \xi_{path}(\gamma)$ is equal to the actual exponent of the nonvanishing asymptotic expansion.

An example described in §12, for structure group $S\ell(3)$, shows that the true exponent ξ is generally smaller than ξ_{path}. This indicates that the expansion cannot be obtained by considering the behaviour of the coefficients in the neighborhood of any one path from P to Q. In particular, the expansion cannot be obtained by multiplying together the expansions for a sequence of nearby points—a global approach is called for instead.

III

My motivation for considering this problem was to try to understand the relationship between two natural but different algebraic structures on the moduli space of representations of the fundamental group of S. These two algebraic structures arise as follows. On one hand, the fundamental group is finitely presented, so a representation can be thought of as a collection of matrices, one for each generator, subject to some equations imposed by the relations. The space of representations is an affine algebraic variety—even after dividing out by the relation of equivalence using geometric invariant theory, which yields the *Betti moduli space* M_B. On the other hand, a representation is the same thing as a flat holomorphic connection on a holomorphic bundle, in other words a first order system of ordinary differential equations on S. The monodromy matrices describing what happens when a solution is continued around a loop give the corresponding representation of the fundamental group. The flat connections on holomorphic bundles are again parametrized by an algebraic variety, the *de Rham moduli space* M_{DR}—here the problem of dividing out by the relation of equivalence is a more complicated problem in geometric invariant theory, treated in [29]. The correspondence between a connection or system of differential equations and its monodromy representation is known as the Riemann-Hilbert correspondence. It gives a complex analytic isomorphism $M_{DR}^{an} \cong M_B^{an}$, but this isomorphism is not algebraic. For example in the case of one dimensional representations, the space of representations of the group is $(\mathbf{C}^*)^{2g}$ while the space of line bundles with holomorphic connection is the universal extension of the g-dimensional Jacobian of the curve by the vector space $\mathbf{C}^g$. The map between these spaces grows exponentially at infinity.

Our parametrized family of systems of equations $dm = (tA+B)m$ corresponds to an algebraic family of connections on the trivial bundle $\mathbf{C}^r$, giving an algebraic curve going out to infinity in M_{DR}. It gives a holomorphic, but not algebraic curve in M_B. If we choose Q to be equal to P and the path to be an element of the fundamental group, then the matrix $m(Q,t)$ is the value of the monodromy representation at that group element. Any invariant polynomial in $m(Q,t)$ gives a coordinate on M_B. We will show in §11 that any polynomial in the matrix entries has an asymptotic expansion. These asymptotic expansions give a description of the behavior of the holomorphic curve in M_B which is the image of an algebraic curve in M_{DR} by the Riemann-Hilbert correspondence.

IV

The methods we use are based on several classical ideas. The first is an iterative series expansion for a solution of a differential equation. The terms in the series are iterated integrals. The iteration expansion is a noncommutative analogue of the power series for e^x. This expansion is ancient. It appears in the work of Liouville [19], but he uses it so casually that it must have been well known at the time; in primitive form it can be traced back to the work of Newton. It was discussed extensively by Picard [26], and has been further studied by Chen [4] and others. In our problem we begin by making a multivalued gauge transformation, and then expand the fundamental solution as a sum of iterated integrals.

The second idea is the method of stationary phase. Classically this method gave asymptotic expansions for integrals like

$$\int_{-\infty}^{\infty} f(x)e^{-tx^2}\,dx.$$

More generally it gives asymptotic expansions for many integrals which have exponentials in t in the integrand. It is very closely connected with the Laplace transform, and we will use this point of view. This theory has been reincarnated in etale cohomology by Laumon [18]. Our §2 below attempts to put Laumon's theory back in terms of complex analysis.

The resulting considerations are related to J. Ecalle's theory of resurgent functions [37], although not touching on its original aspects such as "derivées etrangers". We do not refer to Ecalle's work—our discussion is contained purely in the realm of classical complex analysis.

Each of the integrands in our iteration expansion will contain an exponential in t, so the method of stationary phase gives an asymptotic expansion for each term. In a formal way we could add these up to get an asymptotic expansion for the solution. The main problem is to show that this works, in other words that the expansions for the individual terms add up to an expansion for the sum. The reason for convergence of the perturbation series is the small size of the cycles of integration: the domain of integration for the nth term has size $(1/n!)$. On the other hand, the asymptotic expansions are obtained by moving the cycles of integration. One must show that the cycles remain small after being moved. The problem is that the cycles are in relative homology, so moving them seems to introduce many boundary terms. A simple estimate of the number of boundary terms introduced this way gives at least $(n!)$. Remarkably, enough of these terms are degenerate that the required bounds on sizes of the cycles still hold. The reason for this degeneracy seems to come from the following effect. If one tries to count the number of binary trees with n branch points, one might give a bound by attaching the branches one by one and counting the possibilities at each step, an estimate which comes out to approximately $n!$. However, the number of binary trees is bounded by C^n, reflecting the fact that the order in which the branches are attached is in many cases not relevant. Beyond this main point, it seemed clear that the method would work—the remaining difficulties being mostly technical.

V

There are many obvious directions for further work on this subject. We have made many assumptions which should be removed. The principal example is the assumption that A is diagonal and B has zeros along the diagonal. A sort of reduction to this case can be made using gauge transformations. For example if A is semisimple, then after passing to a ramified cover of X, we can diagonalize A by similarity transformations. Thus if $A_1 = RAR^{-1}$ is diagonal, we consider the equation for the matrix $m_1 = Rm$:

$$dm_1 = tRAR^{-1}m_1 + RBR^{-1}m_1 + (dR)R^{-1}m_1.$$

Now A_1 takes the role of A and $B_1 = RBR^{-1} + (dR)R^{-1}$ takes the role of B. After diagonalizing A we still have the freedom to conjugate by a diagonal matrix, which conjugates B and adds $(dR)R^{-1}$. We may try to choose the diagonal elements of R so as to eliminate the diagonal elements of B. Thus after such gauge transformations we may assume that A is diagonal and B has zeros along the diagonal. The difficulty with this procedure is that the matrix R will be meromorphic instead of holomorphic in the first case, and multivalued in the second case. Thus in order to make this reduction, we must remove the assumption that our connection is holomorphic. A brief discussion is given in §11, but there seems to be essentially new behaviour perhaps not tractable by our methods.

A smaller loose end is the fact that we can only prove that the expansions are nonzero for generic multiples of B. It would be nice to be able to obtain nonzero expansions for all values of B, in other words to show that there is no unwarranted infinite cancellation for special values of B.

VI

Another direction for further research is the question of how to extend the interpretation of the exponent of the asymptotic expansion given for the example of $S\ell(2)$ connections, to connections with larger structure group. As the $S\ell(3)$ example will show, the situation for bigger groups is topologically and geometrically more complicated. It would probably be interesting to try to compute the first terms of the expansion in some specific examples. As a more general problem, one could ask whether the asymptotic expansions for the terms in the sum occur with the expected exponent, namely the smallest real number ξ_{top} such that the cycle of integration exists in the part of the space where the real part of the coefficient of t in the exponential is smaller than ξ_{top}. In order to understand this, one should develop a method for calculating the expansion corresponding to a certain exponent, and then show that the answer is nonzero at the expected exponent.

VII

The method discussed in this monograph might, with suitable modifications, prove to be useful for numerically integrating differential equations of the form we are studying. The fact that the asymptotic expansion has a lower exponent than would become evident from integrating along any particular path means that if the equation is integrated numerically along a path, the size of the error would behave with a higher exponent than the size of the actual answer. This could cause problems for large finite values of t. It is possible that our method of using a series approximation and applying the method of steepest descent to each integral in the approximation could be made into a numerical algorithm. The process of moving the cycle of integration would have to be done concretely. If the cycles of integration could be moved to the point where the supremum of the real part of g was equal to the exponent of the actual answer (as would be the case if the answer to the question posed in the previous paragraph was affirmative), then the sizes of the errors would grow with the correct exponent. Of course this procedure might be too complicated to implement effectively.

VIII

The question we discuss falls under a subject known as *singular perturbation theory*. In that context our system of equations might have been rewritten with $\epsilon = t^{-1}$ as

$$\epsilon dm = (A + \epsilon B)m.$$

This is equivalent to our formulation and we will stick with the notation t.

One of the earliest studies of a singular perturbation problem for ordinary differential equations was by Sturm and Liouville [19] [20] [21] [31]. They studied equations typified by

$$\frac{dK.dL.d\ldots M.dN.du}{dx^{\mu}} + ru = 0,$$

where $K, L, \ldots, M, N$ are positive continuous functions of x. Set $t = r^{1/\mu}$. By introducing new variables $v_1 = u$, $v_2 = t^{-1}N dv_1/dx$, $v_3 = t^{-1}M dv_2/dx$, *etc.*, this equation can be rewritten as

$$\frac{dv}{dx} = t\mathbf{M}v$$

where

$$\mathbf{M} = \begin{pmatrix} 0 & N^{-1} & 0 & \ldots & 0 & 0 \\ 0 & 0 & M^{-1} & \ldots & 0 & 0 \\ & & & \vdots & & \\ 0 & 0 & 0 & \ldots & L^{-1} & 0 \\ 0 & 0 & 0 & \ldots & 0 & K^{-1} \\ -1 & 0 & 0 & \ldots & 0 & 0 \end{pmatrix}.$$

The eigenvalues of $\mathbf{M}$ are the μ different roots $(KL \cdots MN)^{-1/\mu}$. The assumption that the $K, \ldots, N$ are positive implies that the directions of the differences between these roots always remain the same in the complex plane—the roots can not wind around each other. If the functions involved are analytic on the complex plane, this equation could be put into our form by diagonalizing $\mathbf{M}$ with a gauge transformation as described in *V* above. The roots or eigenvalues of $\mathbf{M}$ then become the diagonal entries of A. The matrix B arises because of the gauge transformation. Our method might not apply though, because the resulting matrices will probably have poles. In this sense, the hypotheses discussed in the present work do not subsume those of the earlier works.

Sturm and Liouville asked for asymptotic information about values of r for which there would be a solution u satisfying some linear conditions at the boundary points 0 and 1. At the turn of the century, Birkhoff generalized this boundary value problem, solving it by investigating the asymptotic behaviour of solutions of a family of equations [2] [3]. This suggested the material we will discuss in §12 (see *IX* below). Birkhoff formalized the hypothesis that the roots do not wind around each other with his notion of "region S in the ρ-plane" ([2] p. 220).

Two decades later, in physics, the WKB method of G. Wentzel, H. A. Kramers, and L. Brillouin was a treatment of a singular perturbation problem arising in quantum mechanics.

The next advance was the treatment by Langer of "turning points" of second order equations [17]. In our language, a turning point is a point where the difference between the diagonal entries of A vanishes. These points appear below as critical points of functions g_{ij}. In studies prior to [17], turning points were avoided by hypothesis. Langer's work led to a complicated theory of patching together asymptotic expansions in order to pass through turning points, pursued by Wasow and several other people—see the references in Wasow's book [33]. Other work in the middle of this century included that of Tikhonov [32] and Kruskal [15].

I have tried to find some recent references to mention in the bibliography (although they are not refered to in the body of the paper). These include books by Beals, Deift and Tomei [1], Eckhaus [6], Grasman [10], Hille [11], Mishchenko and Rozov [22], Nayfeh [23], and Wasow [34] as well as the papers of Gingold [9], Nipp [24], and O'Malley [25]. Especially notable seem to be the announcements of Lomov *et. al.* [7] [13] [14]. There are undoubtedly many important works left out of this partial list. Some more references may be found by looking in the bibliographies of the books mentioned.

The references deal almost exclusively with the problem of differential equations on the real line. The first hint of an exception to this principle appears in an example in Wasow's article [35] on "The capriciousness of singular perturbations". His example shows that a turning point in the analytic continuation of a system of equations to the complex plane can affect the behaviour of the solution to the system along the real line.

The principal way in which the present work differs from the previous results refered to above is that our asymptotic expansions are obtained using information about the analytic continuation of a system of equations to a compact Riemann surface. The $S\ell(3)$ example in §11 shows that the exponent of asymptotic behavior is generally smaller than any exponent which could be found by looking at the system of equations along any path. Roughly speaking, it will suffice to piece together solutions along a path only if the roots or diagonal entries of A do not wind past each other in the complex directions. If the roots do wind around, then the correct asymptotic information comes from consideration of the analytic continuation of the system of equations. Our method shows that asymptotic expansions can be obtained even in this winding case.

It has come to my attention more recently that Pham and his students Delabaere and Dillinger have obtained some global results about monodromy of families of second order scalar equations (similar to the rank two vector case) [38]. They applied the theory of resurgent functions of Ecalle for multiplying together asymptotic expansions and keeping track of the lower order terms which occur when higher order terms cancel out.

IX

Sturm and Liouville, and later Birkhoff, were interested in the special values of t for which there would be solutions of a linear boundary value problem. In the last section of this paper §12, we will address this question in our situation. The existence of a solution of the boundary value problem can be expressed as the vanishing of some determinant involving matrix entries of the solution. More generally, suppose P is any polynomial in the matrix coefficients of $m(Q)$. Then P is an entire function of t, whose Laplace transform has locally finite quasi-regular singularities (as will be shown in §11). Under these circumstances, there is an asymptotic expression for $P(t)$ defined and valid in the complement of a branch cut going out from the origin. We will see that one can recover an asymptotic description of the location of the zeros of $P(t)$. Essentially, there are strings of zeros of constant density going to infinity in various directions. The directions are determined by the convex hull of the exponents which occur in the asymptotic expression for $P(t)$.

X

Following Beals, Deift, and Tomei [1], we can think of the function $m(Q, t)$ as the result of a *scattering process*, the scattering refering simply to the ordinary differential equation which causes $m(Q, t)$ to differ from $m(P, t) = I$. In [1] they formulate and treat the inverse scattering problem in their situation. This means, given the scattering data, they try to reconstruct the scattering process. We can formulate the corresponding problem in our situation, as an interesting question for future research.

Problem (Inverse Scattering) *Suppose that a matrix valued function $m(t)$ is given, and suppose one knows that it is the solution to a problem such as posed above. Does $m(t)$ determine the Riemann surface S, the matrices of one forms A (diagonal) and B (with zeros on the diagonal), the point $P \in S$, and the point Q in the universal cover of S, such that $m(t) = m(Q, t)$ is the solution at Q of the differential equation $dm = (tA + B)m$ with initial conditions $m(P) = I$?*

Some immediate qualifications must be made. If $S' \to S$ is a finite ramified covering, and if the data on S is pulled back to S', then the solution on S' will give the same function $m(t)$. Similarly, if we take equations on two Riemann surfaces S_1 and S_2, take the tensor product on $S_1 \times S_2$, and restrict to a curve $S' \subset S_1 \times S_2$, then the solution will not always depend on S'. For example S' could be taken as any curve which passes through fixed points P on $S_1 \times S_2$ and through Q on the universal cover of $S_1 \times S_2$. As long as P and Q are fixed, the solution does not depend on the choice of S'. As a solution to the inverse scattering problem, one might hope for a finite list of such qualifications, and a theorem saying that the data may be otherwise recovered.

A similar question concerns a natural semi-norm for elements of the fundamental group. Fix the Riemann surface S, the system of equations $d - tA - B$, and the base point P. For any element γ in $\pi_1(S, P)$, we can set $Q = \gamma P$ be the translate in the universal cover. Then we obtain a function $m(\gamma, t) = m(\gamma P, t)$. This has an asymptotic expansion for positive real t. If we assume $Tr(A) = 0$ then $det(m(\gamma, t)) = 1$, so the real parts of the exponents of the asymptotic expansion are positive. Let $\xi(\gamma)$ be the real part of the exponent, so

$$\xi(\gamma) = \lim_{t \to \infty} \frac{\log |m(\gamma, t)|}{t}.$$

Clearly $\xi(\gamma\gamma') \leq \xi(\gamma) + \xi(\gamma')$, so ξ is a semi-norm for elements of the group.

Problem *How does $\xi(\gamma)$ compare with the word-length norm on elements of the fundamental group? (The word-length norm is, up to equivalence, independent of the choice of generators.)*

Again certain qualifications must be made as above, for example if the system of equations is pulled back via a map $S \to S'$, then the elements γ in the kernel of $\pi_1(S) \to \pi_1(S')$ have $\xi(\gamma) = 0$. As above, one might ask whether there is a reasonable list of the qualifications which must be made, and a theorem about the remainder of cases.

XI

The main result of this monograph has been announced in [30]. Also treated fully there was the case when B is strictly upper triangular (which eliminates the main convergence problem). That included a description of the method of stationary phase for obtaining the asymptotic expansions of the individual terms in the series.

XII

I would like to thank J. Bernstein for introducing me to the Riemann-Hilbert correspondence; R. Hain for introducing me to the series expansions for the monodromy; and G. Laumon and my father for introducing me to the method of the stationary phase. I would also like to thank W. Thurston, C. McMullen, F. Almgren, N. Elkies, and F. Pham for helpful conversations, and the Department of Mathematics at Princeton University, where this work was done.

1. Ordinary differential equations on a Riemann surface

Let S be a compact Riemann surface. We will consider systems of first order ordinary differential equations on S

$$(d - tA - B)m = 0$$

where A and B are $k \times k$ matrices of holomorphic one-forms on S, t is a complex parameter, and m is a column vector or $k \times k$ matrix of functions. We make the following assumption:

A is a diagonal matrix with one-forms $a_1, \ldots, a_k$ along the diagonal. The diagonal entries of B are equal to zero.

The solutions of the system of differential equations are multivalued holomorphic functions on S, so it is more convenient to introduce the universal cover $Z = \tilde{S}$. This is complex analytically equivalent to a domain in the complex plane, and it is sometimes useful to keep such an embedding in mind.

Fix a base point P in Z (lying above a base point which we also denote by P in S). There is a unique matrix valued solution $m(z)$ defined for $z \in Z$, specified by initial conditions $m(P) = I$. For any point Q on Z, the value $m(Q)$ is well defined. It depends on the parameter t, so we obtain an entire matrix valued function $m(t) = m(Q,t)$ of the complex variable t.

Our aim is to investigate the behavior of $m(t)$ as $t \to \infty$. We can state a theorem which is essentially the main result. Restrict to positive real values of t. Recall that an *asymptotic expansion* for $m(t)$ is an expression

$$m(t) \sim \sum_{i=1}^{r} \sum_{j=J}^{\infty} \sum_{k=0}^{K(j)} c_{ijk} e^{\lambda_i t} t^{-\frac{j}{N}} (\log t)^k$$

where the real parts of the exponents are equal—say $\Re \lambda_i = \xi$ for all i, such that for each M there is a $y(M)$ and a constant $C(M)$ such that for $t \geq y(M)$,

$$\left| m(t) - \sum_{i=1}^{r} \sum_{j=J}^{NM} \sum_{k=0}^{K(j)} c_{ijk} e^{\lambda_i t} t^{-\frac{j}{N}} (\log t)^k \right| \leq C(M) e^{\xi t} t^{-M}.$$

Call the numbers λ_i the *complex exponents* of the expansion, and the number ξ the *real exponent* or just the *exponent*.

Theorem 1 *Given matrices A and B_0 of one forms satisfying the above assumptions, there is a set of numbers $\chi \in \mathbf{C}$ (the complement of a countable set) such that if we let $B = \chi B_0$, then the transport matrix $m(Q,t)$ has a nonzero asymptotic expansion.*

In the remainder of this section we will express the solution m in terms of iterated integrals. We first make a gauge transformation to simplify the problem. Let g_i be the holomorphic function on Z with $g_i(P) = 0$ and $d(g_i) = a_i$,

$$g_i(z) = \int_P^z a_i.$$

Let $E(z,t)$ be the diagonal matrix with diagonal entries e^{-tg_i}, so that

$$d(E) + tEA = 0.$$

Rewrite the original equation as an equation for Em. Note that

$$d(Em) = Ed(m) + d(E)m = E(dm - tAm),$$

so the equation becomes

$$(d - EBE^{-1})Em = 0.$$

Expand the solution Em of this equation as a convergent infinite sum of iterated integrals [21] [26] [4]. Choose a path $\gamma : [0,1] \to Z$ with $\gamma(0) = P$ and $\gamma(1) = Q$. The expansion is

$$E(Q)m(Q) = 1 + \int_{P \le z_1 \le Q} EBE^{-1}(z_1) + \int_{P \le z_1 \le z_2 \le Q} EBE^{-1}(z_2)EBE^{-1}(z_1) + \ldots$$

where the notation $P \le z_1 \le \ldots \le z_n \le Q$ indicates that the integration is to be performed over the cycle

$$\{(\gamma(t_1), \ldots, \gamma(t_n)) : 0 \le t_1 \le \ldots \le t_n \le 1\}.$$

The sum converges for any fixed value of t, because the size of the region of integration is $1/n!$ and the integrand is bounded by C^n. One can see that the sum satisfies the required differential equation by differentiating with respect to Q.

Each integrand in the above sum is a product of matrices, which can be written out as a sum of terms. Let $E(Q,t)m_n(Q,t)$ denote the n-th term in the sum, in other words the n-fold iterated integral. Let $\mathbf{e}_{ij}$ denote the elementary matrix with 1 in the ij entry and zeros elsewhere. Then

$$E(Q,t)m_n(Q,t) =$$

$$\sum_{i_0,\ldots,i_n} \mathbf{e}_{i_n i_0} \int_{P\leq z_1\ldots\leq z_n\leq Q} b_{i_n i_{n-1}}(z_n)\cdots b_{i_1 i_0}(z_1) e^{t(g_{i_{n-1}}(z_n)-g_{i_n}(z_n)+\ldots+g_{i_0}(z_1)-g_{i_1}(z_1))}.$$

Let $g_{ij}(z)$ denote $g_i(z) - g_j(z)$. Introduce the multi-index $I = (i_0, \ldots, i_n)$, and set

$$g_I(z_1, \ldots, z_n) = g_{i_0 i_1}(z_1) + \ldots + g_{i_{n-1} i_n}(z_n) + g_{i_n}(Q).$$

Often useful is the formula

$$g_I(z_1, \ldots, z_n) = \int_P^{z_1} a_{i_0} + \int_{z_1}^{z_2} a_{i_1} + \ldots + \int_{z_n}^{Q} a_{i_n}.$$

Also consider the matrix valued n-form

$$b_I(z_1, \ldots, z_n) = \mathbf{e}_{i_n i_0} b_{i_n i_{n-1}}(z_n) \cdots b_{i_2 i_1}(z_2) b_{i_1 i_0}(z_1).$$

Multiplying the previous equation by $E(Q,t)^{-1}$ and plugging in these notations, we have

$$m(Q,t) = \sum_I m_I(Q,t)$$

where

$$m_I(Q,t) = \int_{P\leq z_1\leq\ldots\leq z_n\leq Q} b_I(z_1, \ldots, z_n) e^{t g_I(z_1,\ldots,z_n)}.$$

Our strategy will be to use the method of stationary phase to obtain asymptotic expansions for the terms $m_n(Q,t) = \sum_{|I|=n} m_I(Q,t)$, then to add these together to obtain an asymptotic expansion for $m(Q,t)$ as $t \to \infty$. The main problem will be to show that the sum of asymptotic expansions converges to an asymptotic expansion for the sum of the functions.

In the next section we will discuss asymptotic expansions for integrals such as m_I (or sums of such integrals). Make the following observations for now. Let $T^{n-1} \subset Z^n$ be the subset of points $(z_1, \ldots, z_n)$ such that either $z_1 = P$, $z_i = z_{i+1}$, or $z_n = Q$. We usually regard P as z_0 and Q as z_{n+1}. T^{n-1} is a union of $n+1$ copies of Z^{n-1}, geometrically the boundary of a tetrahedron. Z and hence Z^n are contractible, and T^n has the homotopy type of an $n-1$-sphere. In particular there is a class σ_{n-1} generating the homology $H_{n-1}(T^{n-1})$, and a class β_n generating the relative homology $H_n(Z^n, T^{n-1})$ with $\partial(\beta_n) = \sigma_{n-1}$. Suppose γ is a path from P to Q, in other words $\gamma(0) = P$ and $\gamma(1) = Q$. Then the chain

$$\{(\gamma(t_1), \ldots, \gamma(t_n)) : 0 \leq t_1 \leq \ldots \leq t_n \leq 1\}$$

is a representative for β_n. The integration in the formula

$$m_I = \int_{\beta_n} b_I e^{tg_I}$$

is taken over this chain. However, the integral only depends on the class of the chain in relative homology. Any two representatives of the same class in $H_n(Z^n, S^n)$ differ by the sum of the boundary of an $n+1$-chain in Z^n and a chain lying inside T^{n-1}. The integrand $b_I e^{tg_I}$ is a holomorphic n-form, so it is closed and it vanishes on the complex $n-1$-dimensional subvariety T^{n-1}. Therefore the integration in the formula for m_I may be taken over any representative for β_n.

The exponential behaviour of the integral is governed by the real part of the function in the exponent, $\Re g_I$. Heuristically the idea is to think of $\Re g_I$ as a Morse function, and to move the cycle of integration β_n down until it just goes over various critical points of $\Re g_I$. Then the asymptotic expansion is a sum of terms coming from these critical points. The name "method of stationary phase" indicates that the nonvanishing contribution is due to the fact that the phase, coming from the imaginary part of g_I, is stationary at a critical point.

We will approach this stationary phase problem through the Laplace transform. This will allow us to retain information about lower order terms in case some higher order terms cancel out. The treatement relies on analytic continuation of the Laplace transform of $m(t)$, obtained by moving cycles of integration around using flows which approximate gradient flows in a suitable sense. The moving of cycles of integration for analytic continuation of the Laplace transform is really the same as moving cycles "down" in the Morse theory picture.

We now mention a variant of Theorem 1 which fixes what is the asymptotic expansion. A sum

$$\sum_n \left(\sum_{i=1}^{r} \sum_{j=J}^{\infty} \sum_{k=0}^{K(j)} c^n_{ijk} e^{\lambda_i t} t^{-\frac{j}{N}} (\log t)^k \right)$$

of asymptotic expressions *converges formally* if the numbers N can be chosen independent of n, if for each i, j, k there are only finitely many nonzero terms c^n_{ijk}, and if there are finitely many exponents λ_i where the supremum $\Re\lambda_i = \xi = \sup_j \Re\lambda_j$ is achieved. In this case, the formal sum of the asymptotic expressions is defined to be equal to

$$\sum_{i=1}^{r} \sum_{j=J}^{\infty} \sum_{k=0}^{K(j)} \left(\sum_n c^n_{ijk}\right) e^{\lambda_i t} t^{-\frac{j}{N}} (\log t)^k$$

where the sum over i is taken over those i such that $\Re\lambda_i = \xi = \sup_j \Re\lambda_j$. If the terms with the highest exponent all cancel out, the formal sum is taken to be the zero expression with real part of the exponent ξ, rather than some nonzero expression with lower exponent.

Variant 1.1 *The terms $m_n(t)$ have nonzero asymptotic expansions. The sum of these expressions converges formally to an expression which is an asymptotic expansion for $m(t)$. For any matrix B_0 we may choose $B = \chi B_0$ so that this expansion for $m(t)$ is not identically zero.*

The individual terms $m^n(Q,t)$ are homogeneous of degree n in B. In principal, for a given choice of B_0 the asymptotic expansions could cancel out and yield the trivial or zero expansion with a given real part ξ of the exponent for $m(t)$. Then the expansion would only come to the statement that $m(t)$ grows more slowly than $t^{-j}e^{\xi t}$ for any j. However, since the parts of the expansion with given degrees of homogeneity in B do not vanish separately, we can choose a generic multiple χB_0 such that the formal sum does not vanish altogether.

2. Laplace transform, asymptotic expansions, and the method of stationary phase

Classically, the method of the stationary phase (or steepest descent) provided asymptotic expansions for integrals such as

$$\int f(z)e^{-tz^2}dz.$$

In this paper we are interested in obtaining asymptotic expansions for more general integrals such as

$$m(t) = \int_\eta be^{tg},$$

where g is a holomorphic function on a complex manifold, b is a holomorphic differential form of top degree, and η is a cycle in homology or relative homology (of real dimension equal to the complex dimension of the manifold). Instead of applying the method of stationary phase directly to such an integral, it will be more useful to take the Laplace transform first. The Laplace transform keeps lower order information which is lost upon going to the asymptotic expansion. If several such integrals are added together and their asymptotic expansions cancel, then an asymptotic expansion at lower exponent can be recovered from the sum of the Laplace transforms.

Suppose that $m(t)$ is an entire holomorphic function of order ≤ 1. This means that there is a bound

$$|m(t)| \leq Ce^{a|t|}.$$

The *Laplace transform* of m is defined to be the integral

$$f(\zeta) = \int_0^\infty m(t)e^{-\zeta t}dt.$$

The integration is taken along a direction in which the integrand is rapidly decreasing. $f(\zeta)$ is defined and holomorphic for $|\zeta| > a$, and it vanishes at ∞. Conversely the function $m(t)$ can be recovered as the *inverse Laplace transform*

$$m(t) = \frac{1}{2\pi i}\oint f(\zeta)e^{\zeta t}d\zeta.$$

Here the path of integration is a large circle running once counterclockwise around the annulus $|\zeta| > a$.

Lemma 2.1 *The operations of taking the Laplace transform of an entire function $m(t)$ of order ≤ 1, and of taking the inverse Laplace transform of a function $f(\zeta)$ defined for $|\zeta| \gg 0$ and vanishing at ∞, are inverses of one another. If*

$$m(t) = \sum_{i=0}^{\infty} a_i t^i$$

then its Laplace transform has power series at infinity

$$f(\zeta) = \sum_{i=0}^{\infty} i! a_i \zeta^{-i-1}.$$

Proof: It is easy to check the formulas for the power series, formulas which then insure that the operations are inverses.

Remark: A rotation of the t plane corresponds to a rotation in the opposite direction in the ζ plane. Thus if $f(\zeta)$ is the Laplace transform of $m(t)$, then $f(te^{-i\theta})$ is the Laplace transform of $m(te^{i\theta})$. We will generally make such rotations implicit in various statements in order to avoid cumbersome notations about directions in the complex plane. In the context of our original problem, a rotation in the t plane could be accomplished by multiplying the matrix A by $e^{i\theta}$.

The behaviour of the function $m(t)$ for large values of t is controlled by the singularities of the Laplace transform (this is a complex analytic version of the corresponding statement about Fourier transform). For example, if ξ is a real number, then $f(\zeta)$ can be analytically continued in the half plane $\Re\zeta > \xi$ if and only if

$$|m(t)| \leq Ce^{t(\xi+\epsilon)}$$

for all $\epsilon > 0$. The infimum of such ξ gives the rate of exponential increase of $m(t)$ for positive real t.

More precise information on the analytic continuation of $f(\zeta)$ leads to more precise asymptotic information for $m(t)$. Fix a so that $f(\zeta)$ is defined for $|\zeta| \geq a$. Say that f has an analytic continuation with *locally finite branching* if for each $M > 0$ there exists a finite subset $S_M \subset \mathbf{C}$, such that if ρ is any path of length less than or equal to M, beginning with $|\rho(0)| \geq a$ and staying within $\mathbf{C} - S_M$, then f can be analytically continued as a holomorphic function along a neighborhood of ρ. This holds particularly if f has an analytic continuation with a discrete set of branch points, but it does not preclude the possibility that the full set of branch points $\bigcup_M S_M$ could be dense. The condition simply says that in order to reach deeper into the set of branch points, the analytic continuations must be taken over long paths winding around earlier branch points.

Suppose that f has an extension with locally finite branching, suppose $s \in S_M$, and suppose ρ is a path of length $\leq M - 2\epsilon$, starting with $|\rho(0)| \geq a$ and leading to $\rho(1) = s \in S_M$, such that $\rho(t)$ is in $\mathbf{C} - S_M$ for $0 \leq t < 1$. Suppose that ρ approaches s along a ray. Let $D^*(s,\epsilon)$ be the punctured disc of radius ϵ about s. For any point in the universal cover of $D^*(s,\epsilon)$ there is a path of length $\leq M$ extending $\rho|_{[0,1-\epsilon]}$ and ending at that point. The function f can be analytically continued along this path to the point on the universal cover of $D^*(s,\epsilon)$. In other words, f can be continued along ρ to a multivalued function on $D^*(s,\epsilon)$.

Say that a multivalued function $f(\zeta)$ on the punctured disc $D^*(s,\epsilon)$ is *regular singular* if it has a convergent power series expansion

$$f(\zeta) = \sum_{j=J}^{\infty} \sum_{k=0}^{K} c_{jk}(\zeta - s)^{\frac{j}{N}} (\log(\zeta - s))^k.$$

Here is a weaker asymptotic notion. A multivalued function $f(\zeta)$ is *quasi-regular singular* at s if there is an expression

$$f(\zeta) \sim \sum_{j=J}^{\infty} \sum_{k=0}^{K(j)} c_{jk}(\zeta - s)^{\frac{j}{N}} (\log(\zeta - s))^k.$$

with the following property. For any n, and any sector in the universal cover of the punctured disc, there is an estimate

$$\left| f(\zeta) - \sum_{j=J}^{n} \sum_{k=0}^{K(j)} c_{jk}(\zeta - s)^{\frac{j}{N}} (\log(\zeta - s))^k \right| \leq C|\zeta - s|^{\frac{n+\epsilon}{N}}$$

with the constant C dependent on n and the sector (note that any $\epsilon < 1$ will do). Let $\hat{f}$ denote the formal power series expansion. The coefficients of the expansion $\hat{f}$ are uniquely determined by f.

Here is a useful criterion for regular singularity.

Lemma 2.2 *Let $T = T_s$ be the monodromy operator on the space of multivalued functions on $D^*(s,\epsilon)$, defined by $Tf(\zeta) = f(s + e^{2\pi i}(\zeta - s))$. Then $f(\zeta)$ is regular singular at s if and only if it has polynomial growth along any angular* *sector in the universal cover of the punctured disc, and if there exist N and K such that*

$$(T^N - I)^K f(\zeta)$$

is single valued.

Proof: If f has polynomial growth and $(T^N - I)f = 0$, then f is a meromorphic function of $(\zeta - s)^{1/N}$, so it has a power series expansion. If $(T^N - I)^{K+1} f = 0$ then we can express $(T^N - I)^K f$ as a power series, multiply by $(\log(\zeta - s))^K$, and subtract from f. The remainder is killed by $(T^N - I)^K$, so we may then proceed inductively.

Remark: What we call regular singular would be refered to in the language of singular points of differential equations, as regular singular with quasi-unipotent monodromy.

Near a singular point we can decompose a function with regular singularities as

$$f = f_{hol} + f_{sing}$$

where $f_{hol} = \sum_{j=0}^{\infty} c_{j0}\zeta^j$ and $f_{sing} = f - f_{hol}$. Note that if $f_{sing} = 0$ then $f = f_{hol}$ can be analytically continued over s. If f has a quasi-regular singularity, then we may divide the power series expression into pieces $\hat{f} = \hat{f}_{hol} + \hat{f}_{sing}$.

Say that our function f has *locally finite regular singularities* if it has a continuation with locally finite branching, and if the multivalued functions on punctured discs around all singular points (s, ρ) are regular singular. Say that f has *locally finite quasi-regular singularities* if it has a continuation with locally finite branching, and if the multivalued functions on punctured discs around all singular points (s, ρ) are quasi-regular singular. Say that f has *faithful expansions* if, at any point s where f is not holomorphic, the expansion $\hat{f}_{sing}$ does not vanish. (In the case of quasi-regular singularities, the possibility that $\hat{f}_{sing}$ might vanish does not seem to be ruled out *a priori*, as it is in the case of regular singularities).

The sum or product of two functions with locally finite (quasi-)regular singularities again has locally finite (quasi-)regular singularities.

The following proposition is the first part of the lemma of stationary phase.

Proposition 2.3 *Suppose that a nonzero function $f(\zeta)$ (which vanishes at infinity) has an extension with locally finite branching. Suppose that f has quasi-regular singularities with faithful expansions. Then the inverse Laplace transform $m(t)$ has a nonzero asymptotic expansion for positive real $t \to \infty$.*

Proof: Recall that

$$m(t) = \frac{1}{2\pi i} \oint f(\zeta) e^{\zeta t} d\zeta.$$

Deform the path of integration until it is a sum of paths which go around critical points with real part ξ and back in the negative real direction from those points, and paths which are supported on points of real part strictly less than ξ:

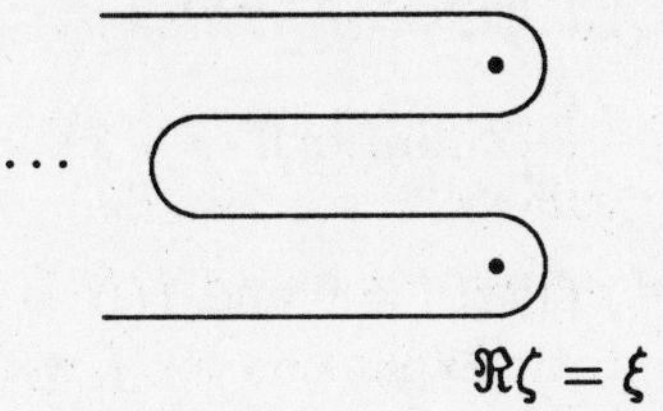

The lower paths will not contribute to the expansion. Let us calculate the contribution from a path σ of the form

Suppose that near the singularity λ_i, $f(\zeta)$ has a quasi-regular series expansion

$$f \sim \sum_{j=J}^{\infty} \sum_{k=0}^{K(j)} c_{ijk} (\zeta - \lambda_i)^{\frac{j}{N}} (\log(\zeta - \lambda_i))^k.$$

Then we will get an asymptotic expansion

$$\int_{\sigma} f(\zeta) e^{\zeta t} d\zeta \sim \sum_{j=J}^{\infty} \sum_{k=0}^{K(j)} e^{\lambda_i t} c_{ijk} \int_{\sigma'} u^{\frac{j}{N}} (\log u)^k e^{ut} du$$

where σ' is the path in the complex plane going from $-\infty$, around the origin, and back to $-\infty$. Make the change of variables $tu = w$. Thus

$$\int_{\sigma'} u^{\frac{j}{N}} (\log u)^k e^{ut} du = \sum_{l=0}^{k} \binom{k}{l} t^{-1-\frac{j}{N}} (-\log t)^{k-l} \int_{\sigma'} w^{\frac{j}{N}} (\log w)^l e^{w} dw.$$

Plugging this in above, and adding up the terms for all λ_i, we get an asymptotic series for $m(t)$. The coefficients are combinations of c_{ijk} and integrals which we denote

$$\Gamma_{j/N,l} \stackrel{def}{=} \int_{\sigma'} w^{\frac{j}{N}} (\log w)^l e^w dw.$$

The trick to dealing with powers of logarithms is that $w^a(\log w)^l = \partial^l w^a / \partial a^l$. The integral becomes a derivative of the Gamma function.

$$\begin{aligned} \Gamma_{j/N,l} &= \frac{\partial^l}{\partial a^l} \int_{\sigma'} w^a e^w dw|_{a=j/N} \\ &= -\frac{\partial^l}{\partial a^l} 2i \sin(\pi a)\Gamma(a+1)|_{a=j/N}. \end{aligned}$$

This coefficient vanishes if and only if $l = 0$ and j/N is an integer greater than or equal to zero. So the asymptotic expansion for f with exponent λ_i vanishes if and only if $\hat{f}_{sing} = 0$.

If the expansion is zero, then by the faithfulness condition, f can be analytically continued across s. If all of the singular points vanish, then f is entire but vanishes at infinity, so $f = 0$.

Remark: By making rotations, the conclusion of the lemma reads more generally that for any angle θ, the function $m(te^{i\theta})$ has a nonzero asymptotic expansion for positive real $t \to \infty$.

Remark: The complex exponents λ_i are exactly the singularities of $f(\zeta)$ with real part ξ equal to the real exponent of asymptotic behaviour of $m(t)$. They are the singularities which are reached first as the half plane $\Re\zeta > \xi$ is moved to the left.

If $m(t)$ is any entire function of order less ≤ 1, define the *convex hull of exponents* $\mathcal{H}(m)$ to be the smallest closed convex subset of C such that for any neighborhood $\mathcal{H} \subset U$,

$$|m(t)| \leq \sup_{h \in U} |e^{ht}|.$$

In terms of the Laplace transform f of m, $\mathcal{H}$ is the smallest closed convex set such that $f(\zeta)$ can be analytically continued to the complement of $\mathcal{H}$. If f has an analytic continuation with locally finite branching, then $\mathcal{H}$ is a polygon. Its corners are at points $s \in S_M$ for $M = 2a$ (where a is a number such that $f(\zeta)$ is defined for $|\zeta| \geq a$). The conclusion of Proposition 2.3 could be rephrased to give an asymptotic expansion for $m(t)$ uniform in a given sector of the complex plane, with exponents on the boundary of $\mathcal{H}$. Each exponent on the boundary of $\mathcal{H}$ contributes an asymptotic series as mentioned above.

If we restrict to positive real values of t, the real exponent of the asymptotic expansion is

$$\xi = \sup_{h\in\mathcal{H}} \Re h.$$

It is the smallest real number such that $f(\zeta)$ can be analytically continued in $\Re\zeta > \xi$, and is also given by the formula

$$\xi = \lim_{t\to\infty} \sup t^{-1} \log |m(t)|.$$

Our application of the method of the stationary phase as outlined above will be to obtain asymptotic expansions for integrals of the form

$$m(t) = \int_\eta be^{tg}$$

where Y is a complex manifold, $Y' \subset Y$ is a closed subvariety, $g : Y \to \mathbf{C}$ is a holomorphic function, b is a holomorphic form of top degree, and η is a representative for a cycle in relative homology $H_i(Y, Y')$, of real dimension $i = dim_{\mathbf{C}} Y$. Notice that be^{tg} is holomorphic and hence closed, and that it restricts to zero on the subvariety Y' of smaller dimension. Thus the integral is independent of the representative for the relative homology class.

The function $m(t)$ is entire of order ≤ 1, and its Laplace transform is given by

$$f(\zeta) = \int_\eta \frac{b}{g-\zeta}.$$

This can be seen by interchanging the order of integration and noting that

$$\int_0^\infty e^{t(g-\zeta)}dt = \frac{1}{g-\zeta}.$$

Let $|\eta|$ denote the support of the representative η. The formula for the Laplace transform $f(\zeta)$ is valid for any ζ in that connected component of the complement of $g_*|\eta|$ which contains ∞.

Our goal is to obtain the analytic continuation of $f(\zeta)$ by moving the cycle of integration η. The following lemma shows how to accomplish this in a simple case. It will not be directly relevant for our purposes, but the proof in this case should serve as a guide to our more complicated argument.

Proposition 2.4 *Suppose the function g is proper. Then $f(\zeta)$ has a continuation with locally finite regular singularities. Consequently, the function $m(t)$ has a nonzero asymptotic expansion for positive real $t \to \infty$ (unless $m(t) = 0$).*

Proof: Define the notion of critical point as follows. A point y is *non-critical* for g on (Y, Y') if locally near y the triple (Y, Y', g) is diffeomorphic to a product $(Y_0 \times D, Y_0' \times D, pr_2)$ where (Y_0, Y_0') is the fiber of g through y, and D is a small disc around $g(y)$ in $\mathbf{C}$. A point is a *critical point* if it is not non-critical. This condition can be understood as follows. Choose a stratification $Y' = \bigcup Y_\alpha$ so that along each stratum Y_α, the pair (Y, Y') is differentiably the product of the stratum and the cross-section. In particular the strata are smooth. Such a stratification exists, and is called a Whitney stratification. The open set $Z - S$ is considered as the biggest stratum. The critical points are points y such that $dg|_{Y_\alpha}(y) = 0$ where Y_α is the stratum containing y.

Since g is proper, the set of images of critical points is a discrete subset $S \subset \mathbf{C}$. If ρ is a path in $\mathbf{C} - S$ then there is a a neighborhood U of ρ such that the inverse image $g^{-1}(U) \subset (Y, Y')$ is diffeomorphic to a product $(Y_0, Y_0') \times U$. Suppose η is a class in relative homology $H_n(Y, Y')$ such that the image in $\mathbf{C}$ of the support of η does not meet $\rho(0)$. Then the triviality of (Y, Y') over U implies that there is a homology φ whose support does not meet $\rho(0)$, such that $\partial\varphi = \eta' - \eta$ up to boundary terms in Y', and such that the support of η' does not meet ρ. This gives an analytic continuation of the integral

$$f(\zeta) = \int_\eta \frac{b}{g - \zeta} = \int_{\eta'} \frac{b}{g - \zeta}$$

from ζ in a neighborhood of $\rho(0)$, to ζ in a neighborhood of ρ. This shows that $f(\zeta)$ has an analytic continuation with branch points in the discrete set $S \subset \mathbf{C}$. This gives locally finite branching, by letting S_M consist of the set of all singularities within a distance M from the circle $|\zeta| = a$.

Now we have to prove that f has regular singularities. Choose a path to a singular point, and apply the above process of retraction along that path. We obtain a multivalued function on the punctured neighborhook of the singularity. Assume for simplicity that the singularity is at $x = 0$, and that there are no other singularities in the unit disc $\{|x| \leq 1\}$. Assume that the representative for the cycle of integration does not meet a neighborhood of the fiber $g^{-1}(1/2)$. There is a retraction from the inverse image of the complement of a small disc $g^{-1}(\{|x - 1/2| \geq \epsilon\})$ to the set

$$g^{-1}(\{|x| \geq 1\} \cup [-1, 0]),$$

the union of inverse images of the complement of the unit disc and a line segment going to the origin. Apply this retraction to the cycle of integration, to obtain a new cycle of integration which can be written

$$\eta = \eta_1 + \eta_2 + \eta_3,$$

with η_1 supported in $g^{-1}(\{|x| \geq 1\})$, η_3 supported in $g^{-1}(0)$, and

$$\eta_2 = \bigcup_{x \in [-1,0]} \nu_x.$$

Here ν_x are a collection of homologous $(n-1)$-cycles in the fibers $g^{-1}(x)$ respectively. They fit together, for x in the segment $[-1, 0]$, into the cycle η_2. In order to make the retraction and decompose η in this way, it may be convenient to resolve singularities of (Y, Y', g) first. After resolving, the singularities have nice normal forms, and it is easy to see how to do the retraction (see [5] for a thorough discussion). Then by projecting back to the original (Y, Y') we obtain the decomposition as desired.

The integral f decomposes as a sum $f = f_1 + f_2 + f_3$. The term $f_1(\zeta) = \int_{\eta_1} b/(g - \zeta)$ is holomorphic across the origin, so we can ignore it. The term $f_3(\zeta) = \int_{\eta_3} b/(g - \zeta)$ is equal to zero, since a holomorphic differential form β restricts to zero on a subvariety $g^{-1}(0)$ of smaller dimension. We have to prove that $f_2(\zeta) = \int_{\eta_2} b/(g - \zeta)$ is regular singular.

The map $\wedge dg : \Omega^{n-1}_{Z/D} \to \Omega^n_Z$ is an isomorphism outside of the critical point of g (it fails to be an isomorphism only if g has a critical point on Z). Thus we may write $b = c \wedge dg$ with c a section of $\Omega^{n-1}_{Z/D}$, possibly meromorphic at the origin. Our integral is now

$$f_2(\zeta) = \int_{x=-1}^{0} \frac{1}{x - \zeta} \left(\int_{\nu_x} c \right) dx.$$

The cycle ν_x is an element of $H_{n-1}(Y_x, Y'_x)$. It can be extended to a multivalued function of x with values in that relative homology group. The function

$$a(x) = \int_{\nu_x} c$$

becomes a multivalued analytic function of x on $D^*(0, 1)$, and

$$f_2(\zeta) = \int_{-1}^{0} \frac{a(x)}{x - \zeta} dx.$$

From the resolution of singularities, one can see that the sizes of the cycles $\nu(x)$ are bounded polynomially in $|x|$, so $a(x)$ grows at most polynomially in $|x|$. In fact, the cycle η_2 has finite volume, so the integral $f_2(\zeta)$ is absolutely convergent, so $|a(x)|$ is smaller than $|x|^{-1}$.

We can remove the multiplicity of values of a in the following way. Choose a multivalued basis $\gamma_1(x), \ldots, \gamma_k(x)$ for the relative homology $H_{n-1}(Y_x, Y_x')$. Set $a_i(x) = \int_{\gamma_i(x)} c$. It is a multi-valued vector of functions, bounded polynomially in $|x|^{-1}$ (again this can be seen by resolution of singularities). If $\nu_x = \sum v_i\gamma_i(x)$ (the coefficients v_i being single valued), then $a(x) = \sum v_i a_i(x)$. Let M denote the monodromy matrix expressing the transformation undergone by γ_i upon continuation around the origin, let $\log M$ be a logarithm for M (divided by $2\pi i$), and set

$$P(x) = e^{\log M \log(x)}.$$

It is a multivalued analytic matrix function of x chosen so that the vector $h_i(x) = \sum P_{ij}(x)a_j(x)$ becomes a single valued function of x. Now

$$a(x) = \sum v_i P_{ij}^{-1} h_j(x).$$

The bound on $a_j(x)$ implies that $h_i(x)$ are meromorphic. The Laurent expansions for $h_i(x)$ translate to an expansion

$$a(x) = \sum_{j=J}^{\infty} \sum_{k=0}^{K-1} c_{jk} x^{\frac{j}{N}} (\log x)^k.$$

Here the numbers K and N come from the quasi-unipotence of the monodromy matrix: $(M^N - I)^K = 0$ [16]. Since b is absolutely integrable over ν_2, $a(x)dx$ must be absolutely integrable over the path from -1 to 0. Therefore $J > -N$, in other words there are no terms with exponent $\frac{j}{N} \leq -1$. From this it is easy to see that $f_2(\zeta)$ has polynomial growth as $\zeta \to 0$. In order to prove that $f_2(\zeta)$ is regular singular, we must show that $(T^N - I)^K f_2$ is single valued. This is due to the quasi-unipotence of the monodromy matrix, which causes $(T^N - I)^K a = 0$. Make the substitution $\zeta = z^N$. Then

$$\begin{aligned} f(z) &= \int \frac{a(x^N)}{x^N - z^N} d(x^N) \\ &= \sum_{\mu^N=1} \int c_\mu \frac{a(x^N)}{x - \mu z} N x^{N-1} dx \end{aligned}$$

by partial fractions. By using a new $a(x)$ as indicated by this formula, we can reduce to the case $N = 1$, and suppose

$$f(\zeta) = \int \frac{a(x)}{x - \zeta} dx$$

with $(T - I)^K a = 0$ this time. Tf is given by the same formula, but with the path of integration going on the other side of ζ. Thus $(T - I)f$ is equal to the residue of the integrand at ζ, which is

$$(T - I)f(\zeta) = 2\pi i a(\zeta).$$

Since $(T - I)^K a$ is single valued, $(T - I)^{K+1} f$ is single valued. Thus f is regular singular. This proves the proposition.

Remark: The estimate for the power of logarithms K in f is one greater than the corresponding value for a.

Infinite sums

Next we will give some conditions on an infinite sum of functions f_n with locally finite regular sigularities intended to insure that the sum

$$f(\zeta) = \sum_{n=1}^{\infty} f_n(\zeta)$$

exists and is a function with locally finite branching, quasi-regular singularities, and faithful expansions.

2.5.0: There is an a such that the sum is uniformly convergent for $|\zeta| \geq a$.

2.5.1: There are finite subsets S_M which work for all f_n in the definition of locally finite branching.

2.5.2: For any piecewise linear path ρ of length $\leq M$ in $\mathbf{C} - S_M$, with $|\rho(0)| \geq a$, there is a neighborhood U of ρ such that the continuations satisfy

$$|f_n^\rho(\zeta)| \leq C^n n^{-\epsilon n}$$

for $\zeta \in U$. In particular the sum of the analytic continuations f_n^ρ is uniformly convergent on U.

2.5.3: Fix a singular point s and an angular sector U in a small punctured disc around s. Then there is a number β, an $\epsilon > 0$, and a constant C such that for $0 \leq k \leq \epsilon n - \beta$, we have

$$\left|\frac{1}{k!}\frac{d^k}{d\zeta^k}f_n(\zeta)\right| \leq C^n n^{-\epsilon n}$$

for ζ in the angular sector U. The constants C and ϵ are uniform in n for any fixed value of k.

2.5.4: The terms $f_n(\zeta)$ have locally finite regular singularities and there is a uniform bound for the number N which occurs in the expansions for the f_n.

2.5.5: There are independent sub-**Q**-vector spaces $H_n \subset \mathbf{C}$ (in other words $\bigoplus H_n \hookrightarrow \mathbf{C}$) such that the coefficients c^n_{jk} of the singular parts of the expansions $\hat{f}_{n,sing}(\zeta)$ are contained in H_n.

Proposition 2.5 *The conditions (2.5.0), (2.5.1), and (2.5.2) imply that f has an extension with locally finite branching. Conditions (2.5.3) and (2.5.4) imply that at any singular point, f has a quasi-regular singularity. Condition (2.5.5) implies that f has faithful expansions.*

Proof: Condition (0) implies that $f(\zeta)$ is defined and holomorphic for $|\zeta| > R$. Conditions (1) and (2) clearly imply that f has an extension with locally finite branching, the singularities for paths of length $\leq M$ comprising the set S_M.

Apply conditions (2.5.3) and (2.5.4), near a singular point s. Fix an angular sector U in a small punctured disc around s. For brevity, use the notation $f_n^{(k)} = d^k f_n/d\zeta^k$ for derivatives. The limiting values of the derivatives $f_n^{(k)}(s)$ are well defined for $0 \leq k \leq \epsilon n - \beta - 1$, because of the bound (2.5.3) and the formula

$$f_n^{(k)}(s) = f_n^{(k)}(s') + \int_{s'}^{s} f_n^{(k+1)}(t)dt.$$

In fact, the derivatives at s satisfy the same type of bound as in (2.5.3). Fix a number L, and choose n_0 so that $\epsilon n_0 - \beta > L$. Write $f = f_{\leq n_0} + f_{>n_0}$ with

$$f_{\leq n_0}(\zeta) = \sum_{n=0}^{n_0} f_n(\zeta)$$

and

$$f_{>n_0}(\zeta) = \sum_{n>n_0} f_n(\zeta).$$

By assumption (2.5.4) the terms f_n for $n \le n_0$ are regular singular, so there is a Laurent polynomial $\hat{f}^L_{\le n_0}(\zeta)$ in $\zeta^{1/N}$ and $\log(\zeta)$, with

$$|f_{\le n_0}(\zeta) - \hat{f}^L_{\le n_0}(\zeta)| \le C|(\zeta - s)|^L$$

uniformly for $\zeta \in U$. On the other hand, for $n > n_0$ set

$$\hat{f}^L_n(\zeta) = \sum_{k=0}^{L-1} \frac{f_n^{(k)}(s)}{k!}(\zeta - s)^k.$$

The bound of type (2.5.3) for the derivatives at s shows that the sum of polynomials of degree L,

$$\hat{f}^L_{>n_0}(\zeta) = \sum_{n>n_0} \hat{f}^L_n(\zeta)$$

converges. Let $R^L_n(\zeta) = f_n(\zeta) - \hat{f}^L_n(\zeta)$. By the Taylor remainder formula

$$R^L_n(\zeta) = \int_s^\zeta f_n^{(L)}(x)\frac{(x-s)^{L-1}}{(L-1)!}dx$$

and the bound (2.5.3), we have that for $n > n_0$

$$|R^L_n(\zeta)| \le C^n n^{-\epsilon n}|(\zeta - s)|^L$$

uniformly for $\zeta \in U$. Thus the sum of the remainders converges with a bound of $C|(\zeta - s)|^L$. We may set $\hat{f}^L = \hat{f}^L_{\le n_0} + \hat{f}^L_{>n_0}$, then

$$|f(\zeta) - \hat{f}^L(\zeta)| \le C|(\zeta - s)|^L$$

for $\zeta \in U$. Since this works for any L, it shows that f has quasi-regular singularities.

Finally we show that it has faithful expansions, using condition (2.5.5). From the above argument, the singular part of the expansion $\hat{f}_{sing}$ is the formal sum of the singular parts $\hat{f}_{n,sing}$ of the individual terms in the sum. Suppose that $\hat{f}_{sing} = 0$. Under condition (2.5.5), the coefficients of the terms lie in independent $\mathbf{Q}$-vector spaces in $\mathbf{C}$, so the only way the formal sum can vanish is that $\hat{f}_{n,sing}$ vanishes for all n. Since the f_n have regular singularities, this means that $f_n(\zeta)$ are holomorphic across $\zeta = s$. In particular they are single valued functions, and we have bounds of the form (2.5.2) on the boundary of a small disc around s. Cauchy's formula gives the same bounds uniformly near

s, so the sum $f = \sum f_n$ converges uniformly on a small disc around s, giving the continuation of f across s. This completes the proof of the proposition.

Remark: If the sum satisfies conditions (2.5.0) – (2.5.4), then the sum of the expansions for $f_{n,sing}(\zeta)$ converges formally to an expansion for $f_{sing}(\zeta)$. But the formal sum could vanish even if the individual expansions are nonvanishing.

3. Construction of flows

In this section we construct some flows on the one dimensional manifold Z. These will be used in following sections to move relative homology cycles. We will take some care in the construction of the flows, to obtain technically useful properties.

Suppose that g is a holomorphic function on Z, such as one of the functions $g_{ij}(z) = g_i(z) - g_j(z)$. We want to construct a flow $f(z,t)$ with the property that $f(z,0) = z$, and $g(f(z,t))$ is "downwind" of $g(z)$ in a certain desired direction. In other words, the time derivative of $g(f(z,t))$ is contained in an angular sector of the form

$$S(\pm\delta) \stackrel{def}{=} \{re^{i\theta} : \theta \in [\pi - \delta, \pi + \delta]\},$$

so $g(f(z,t))$ is contained in an angular sector of the form

$$S(g(z), \pm\delta) \stackrel{def}{=} \{g(z) + re^{i\theta} : \theta \in [\pi - \delta, \pi + \delta]\}.$$

We would also like to insure that at $t = 1$, the flow has the effect of moving $g(f(z,t))$ a certain distance away from $g(z)$. This will be possible unless critical points of g are encountered first. We require some special behaviour as the flow moves past critical points. There will be a one dimensional subset $\Lambda \subset \mathbf{C}$, the union of paths which are approximately paths of steepest descent leading away from critical points of g. The flow f will have the effect of moving points to Λ, and then along Λ away from the critical points.

Recall that we are admitting the possibility of rotating the t or ζ planes. This is the same as multiplying the function g by $e^{i\theta}$. After making such a rotation, we can assume that the desired direction of flow is in the negative real direction. Note that $g(P) = 0$ for any of the functions g_{ij} considered. Thus rotation preserves $g(P)$.

Our construction of flows will make reference to four numbers, a choice of angular error δ, a choice of small number σ, a choice of number L, and a choice of radius R. The number L represents the minimum amount by which the real part of g should be decreased by the flow, unless a critical point is encountered. The angular error represents the maximum allowed deviation from the negative real direction, for the direction in which $g(z)$ moves when z is moved by the flow. The σ is a small number which indicates what happens when the flow goes past a critical point.

Assume that an embedding of Z in $\mathbf{R}^2$ is fixed, and use notions of linear and polynomial maps with respect to this embedding. Fix a complete metric on Z, and measure distance $d(x,y)$ using this metric unless otherwise indicated. The radius R indicates the region $d(z,P) \leq R$ in which we force certain properties to hold. We will construct a function

$$f : Z \times [0,1] \to Z$$

and a one-dimensional subset (or zero dimensional in case $dg = 0$)

$$\Lambda \subset Z$$

with a collection of properties. We will first describe the constructions of the flows, and then state the properties.

The case $dg = 0$ is the special case of the flows to be labeled $f^0(z,t)$ below. The fixed flow f^0 does not depend on rotations of g, and is to be kept the same and used throughout our several applications of the procedures outlined below (because our inductive hypotheses will refer to it). If we assume that the image of the embedding $Z \subset \mathbf{R}^2$ is a convex set, then choose a point $\Lambda^0 \in Z$ and define

$$f^0(z,t) = \Lambda^0 + (1-t)(z - \Lambda^0)$$

using the linear structure of $\mathbf{R}^2$. It contracts Z to the point Λ^0 in unit time.

The construction in case $dg \neq 0$ is slightly more complicated. Here is the general *ansatz.* Divide Z into a union of polygons U_i, separated by straight edges. In each polygon, choose a constant vector field V_i, and try to define $f(x,t)$ to be the flow along the resulting piecewise linear vector field V. The subset Λ will be a union of some edges of the polygons. Let Z^Λ denote the space resulting from cutting along Λ (and inserting the pre-image of Λ as boundary). The following proposition provides a preliminary version of the flow $f_1(z,t)$, which later will be modified along Λ to give $f(z,t)$.

Proposition 3.1 *Assume that the piecewise constant vector field V is bounded with respect to some complete metric on Z. Assume that there are an open covering $\{U_\alpha\}$ of Z^Λ and continuous piecewise linear embeddings $\varphi_\alpha : U_\alpha \hookrightarrow \mathbf{R}^2$, with the following properties. U_α is mapped to an open neighborhood of the origin, or to a neighborhood of the origin in the half-plane $x_1 \geq 0$. In the latter case, the boundary $\Lambda \cap U_\alpha$ is mapped into the boundary line $x_1 = 0$. In either case, suppose*

$$(d\varphi_\alpha)_*(V) = \frac{\partial}{\partial x_1}.$$

Then there are continuous piecewise linear functions $f_1(z,t) : Z \times \mathbf{R}^+ \to Z$ *and* $\tau(z) : Z \to \mathbf{R}^+$ *such that* $f_1(z,t)$ *is not in* Λ *for* $t < \tau(z)$, $f_1(z,t) = f_1(z,\tau(z)) \in \Lambda$ *for* $t \geq \tau(z)$, *and such that*

$$\frac{\partial f_1(z,t)}{\partial t} = V(f_1(z,t))$$

for $t < \tau(z)$. *This equation may be interpreted precisely by saying that* $\partial\varphi_\alpha f_1(z,t)/\partial t$ *is equal to* $\partial/\partial x_1$ *for* $f_1(z,t) \in U_\alpha$.

Proof: By patching together flows defined on open sets, we obtain an open neighborhood U of $Z \times \{0\}$ in $Z \times [0,1]$, a function $\tau_0(z) \in \mathbf{R}$, and a map $f_1 : U \to Z$ such that

$$\frac{\partial f_1(z,t)}{\partial t} = V(f_1(z,t))$$

for $f_1(z,t) \in U$ and $t \leq \tau_0(z)$. If $t \geq \tau_0(z)$ and $(z,t) \in U$ then $f_1(z,t) = f_1(z,\tau_0(z))$. In particular, if $s,t \geq 0$ then $f_1(f_1(z,s),t) = f_1(z,s+t)$. Suppose $Z_1 \subset Z$ is a relatively compact subset and T is a number. Since the vector field is bounded with respect to a complete metric, there is a relatively compact neighborhood Z_2 such that if f_1 is any partially defined flow satisfying the conditions of the proposition, and $z \in Z_1$, $t \leq T$, then $f_1(z,t) \in Z_2$. We can choose μ such that if $z \in Z_2$ and $0 \leq t \leq \mu$ then $(z,t) \in U$. For any $z \in Z_1$, and $t \in [a\mu,(a+1)\mu]$ with $t \leq T$, define inductively $f_1(z,t) = f_1(f_1(z,a\mu),t)$. This works for any Z_1 and any T, and the choices all agree where relevant. This gives the function $f_1(z,t)$. We can let $\tau(z)$ be the smallest time such that $f_1(z,\tau(z)) \in \Lambda$.

We would like to understand when the local criteria are satisfied. This only depends on the directions of the vector fields. There are several possibilities for the relationship between a given edge and the two vector fields on either side:

in − *in* :

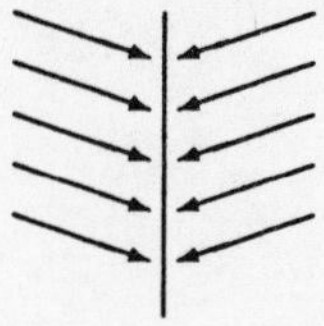

$in - out$:

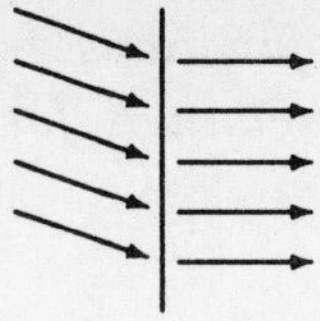

$out - out$:

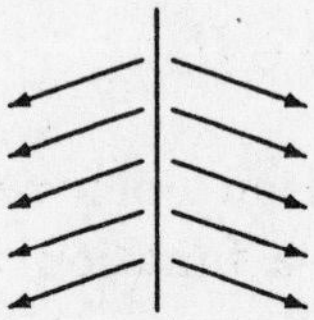

We assume that the vector fields are chosen generically so that they are never tangent to an edge. Of the three possibilities pictured above, only the last ($out - out$) causes problems for continuity. We will arrange the construction so that the $out - out$ picture never occurs. At an $in - out$ edge, a continuous piecewise linear transformation could be made so that the vector fields would be the same on both sides, thus Proposition 3.1 applies. On the other hand, the subset Λ must consist exactly of the union of the $in - in$ edges.

There is still some question of what happens at a vertex. We will make a special construction near vertices to be placed at critical points of g. We will assume that the remainder of the vertices have three edges, with picture like

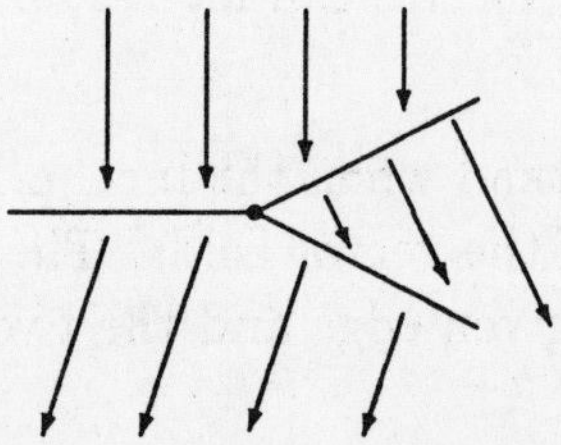

or, in the case of a vertex on Λ,

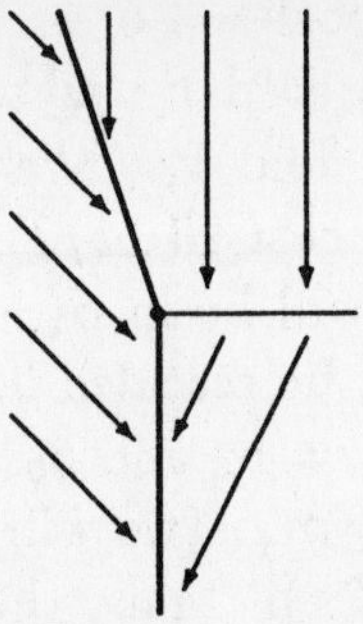

These pictures satisfy the local requirements of the proposition (the second one must be cut along Λ as discussed above). In the first picture, for example, the piecewise linear transformation may be made after making further divisions along the flow lines which go into and out of the vertex. It is left as an exercise to write down the transformations φ_α in these cases.

Along Λ we can make a further choice. Choose a continuous piecewise polynomial function

$$f_\Lambda(\lambda, t) : \Lambda \times [0,1] \to \Lambda$$

with $f_\Lambda(\lambda, 0) = \lambda$. Recall that $\tau(z)$ was the first time when $f_1(z,\tau) \in \Lambda$. Let $\lambda(z) = f_1(z, \tau(z)) \in \Lambda$. Under the hypotheses of the proposition, $\tau(z)$ and $\lambda(z)$ are continuous piecewise linear functions of z. We may alter the definition of the flow for times $t \geq \tau(z)$ in the following way. For $t \leq \tau(z)$, keep $f(z,t) = f_1(z,t)$ as defined in the proposition. For $t \geq \tau(z)$, set

$$f(z,t) = f_\Lambda(\lambda(z), t - \tau(z)).$$

This provides our final version of $f(z,t)$.

Now we turn to the specific situation related to a function $g = g_{ij}$ with $dg \neq 0$, and describe the construction of the set Λ, the vector field V, and the flow f_Λ.

Choose the linear structure $Z \subset \mathbf{R}^2 \cong \mathbf{C}$ to be holomorphic near critical points of any $g = g_{ij}$. Suppose that s is a critical point of g, with local holomorphic coordinate u which is linear ($u(s) = 0$). Let $m \geq 1$ denote the multiplicity of zero of $dg(s)$. We can write

$$dg = ae^{i\omega}u^m du + \ldots$$

with $a \in \mathbf{R}^+$. Let $\chi = ae^{i\omega}u^m du$ be the leading term. We can ignore the higher order terms, for they can be made arbitrarily small relative to the leading term, by restricting to a small neighborhood of s. We will divide a neighborhood of s into angular sectors U_i separated by rays, and in each angular sector we will specify a constant vector field V_i. We will construct these so that for $z \in U_i$, $\chi(z)_*(V_i)$ lies in the sector $S(\pm\delta/2)$. Then by restricting to a small neighborhood of s, $dg(z)_*(V_i)$ will lie in $S(\pm\delta)$. This condition depends only on the direction of V_i, so it can be rescaled later.

Choose sequences $\alpha_0, \ldots, \alpha_n = \alpha_0$ and $\beta_0, \ldots, \beta_n = \beta_0$ of points on the unit circle, so that $\beta_{i-1} \leq \alpha_i \leq \beta_i \leq \alpha_{i+1}$ (we will confuse points on the unit circle with their arguments modulo 2π). Make the points very close together, say $|\beta_i - \beta_{i-1}| \leq \delta/2m$. Let

$$U_k = \{z = re^{i\theta} : \theta \in [\beta_{k-1}, \beta_k]\}$$

and set $V_k = \Re(e^{i(\pi-\omega-m\alpha_k)}\partial/\partial u)$. Then if $z = re^{i\theta} \in U_k$,

$$\chi(z)_*(V_k) = ar^m e^{i(\omega+m\theta)} e^{i(\pi-\omega-im\alpha_k)} = ar^m e^{i(\pi+m(\theta-\alpha_k))}.$$

Since θ and α_k are in the interval $[\beta_{i-1}, \beta_i]$ of length $\leq \delta/2m$, we have $\pi + m(\theta - \alpha_k) \in [\pi - \delta/2, \pi + \delta/2]$, so $\chi(z)_*(V_k)$ lies in the sector $S(\pm\delta/2)$.

The remaining problem is to see what types of edges we have produced (the reader might draw a picture). They will mostly be *in* $-$ *out* edges, with possible exceptions near values of θ with $\pi - \omega - m\theta = \theta$ or $\pi - \omega - m\theta = \theta + \pi$. In this first case the vector field points away from the origin, in the second case it points towards the origin. Choose the angles α and β so that all solutions in the first case become edge angles β, and all solutions in the second case become (strictly) interior angles α. An edge corresponding to angle β_k with $\pi - \omega - m\beta_k = \beta_k$ becomes an *in* $-$ *in* edge, to be included in the set Λ. On the clockwise side of β_k the vector field is determined by α_k which is slightly smaller than β_k. Thus the vector field is at an angle $\pi - \omega - m\alpha_k$ slightly larger than β_k, so it points in toward the β_k edge. The same happens on the other side. The edges of the sector containing an α_k with $\pi - \omega - m\alpha_k = \alpha_k + \pi$ become *in* $-$ *out* edges. The vector field in U_k points at angle $\pi - \omega - m\alpha_k = \alpha_k + \pi$, toward the origin. Thus it points into the two edges of that sector. In the next sector counterclockwise U_{k+1}, α_{k+1} is slightly larger than α_k, so the angle $\pi - \omega - m\alpha_{k+1}$ is slightly clockwise from $\alpha_k + \pi$. Since the (reverse) direction of the edge between U_k and U_{k+1} is slightly counterclockwise from angle $\alpha_k + \pi$, this makes an *in* $-$ *out* edge. The other edges are all *in* $-$ *out* edges.

There are $m + 1$ *in* $-$ *in* edges emanating from the critical point (approximating paths of steepest descent for the function $\Re g$). These will be included as beginnings of branches of Λ. If the local picture is cut along these edges, it falls into $m+1$ pieces. In each of these pieces, one can see by looking at the picture that the flow retracts a neighborhood of Λ onto Λ, satisfying the local smoothability hypotheses of Proposition 3.1 (one can use for piecewise linear coordinates, the time $\tau(z)$ until the flow hits Λ, and the position $\lambda(z)$ in Λ first hit).

Choose small polygons around the critical points of g, and use the local picture described above to define the piecewise constant vector field V there. Extend each branch of Λ started near a critical point into a piecewise linear path going to infinity in Z, approximately in the direction of the negative gradient of $\Re g$. Choose piecewise linear coordinates λ along the branches of Λ, with $\lambda = 0$ at the critical points and λ increasing monotonically away from the critical points. Choose the direction of Λ so that $\partial g(z(\lambda))/\partial\lambda$ lies in the sector $S(\pm\delta/2)$. This will be compatible with the choices made above if we restrict the initial segments of the branches defined near critical points, to small polygonal neighborhoods of the critical points. Use the slight freedom of direction in choosing Λ to insure that a given branch of Λ does not pass through any of the polygonal neighborhoods of other critical points. The polygonal neighborhoods may have to be made smaller again. Finally, divide up the remainder of Z into sufficiently small polygonal pieces U_i, in such a way that the vertices have no more than three edges, and such that the edges are sufficiently transverse to the gradient of $\Re g$. If v is a tangent vector to an edge other than an edge of Λ or an edge near a critical point constructed above, then $dg_*(v)$ should make an angle of at least $10°$ with the real axis. By making the polygonal pieces sufficiently small, one can insure that dg points in approximately the same direction throughout a polygonal piece. Choose constant vector fields V_i so that $dg_*(V_i)$ points in the direction of the sector $S(\pm\delta)$. We may assume (using the extra freedom of $\delta/2$) that in any piece adjacent to Λ, the vector field points towards Λ. Any edge not in Λ is an *in – out* edge, because on either side, the vector field points along the negative gradient of $\Re g$, and the edge is sufficiently transverse to this gradient. Similarly, at the triple vertices, the picture is the one shown previously. At vertices in Λ the picture is like the other one shown above. The local hypotheses of the proposition are satisfied.

In order to apply the proposition to obtain a flow, we need to insure that the size of the vector field remains sufficiently bounded. None of our constructions have depended on the lengths of the vector fields, so the constant vector fields far away from the origin may be rescaled to become sufficiently small. We will demand certain properties below, but always in a compact region, to leave room for the rescaling at large distances. Proposition 3.1 will apply to yield a flow $f_1(x,t)$.

We now define the function $f_\Lambda : \Lambda \times [0,1] \to \Lambda$ as follows. It preserves the branches of Λ coming from the critical points. On each branch, we can use the monotone coordinate λ to define the function:

$$f_\Lambda(\lambda, t) = (1 + C_1 t)\lambda.$$

The constant C_1 will be chosen below. Putting this together with our flow as defined above, we get a function $f(z,t) : Z \times [0,1] \to Z$ defined by the flow $f_1(z,t)$ for $t \leq \tau(z)$, and defined by $f(z,t) = f_\Lambda(\lambda(z), t - \tau(z))$ for $t \geq \tau$.

The next task is to see what properties the flow has, and how to choose multiples of the vector field and how to choose the constant C_1 above, so as to force good properties in the region $d(z,P) \leq R$. These will refer to the case $(dg \neq 0)$, the properties in the simpler case $(dg = 0)$ being clear.

3.2 *Properties of* Λ:

3.2.1 Λ is a graph which decomposes as a finite disjoint union of connected components indexed by the set of critical points ℓ:

$$\Lambda = \bigcup_{\ell} \Lambda(\ell).$$

3.2.2 $\Lambda(\ell) - \{\ell\}$ is a disjoint union of continuous piecewise linear curves. There is a piecewise linear coordinate λ which is equal to 0 at ℓ and which increases monotonically going away from ℓ on each of the branches of $\Lambda(\ell)$.

3.2.3 Moving out along a branch of Λ, the value of g moves in approximately the negative real direction. In other words, $\partial g(z(\lambda))\partial\lambda \in S(\pm\delta)$. Thus if $\lambda(z) \geq \lambda(y)$ then $g(z) \in S(g(y), \pm\delta)$. In particular, $\Re g(\ell) = \sup_{z\in\Lambda(\ell)} \Re g(z)$.

3.2.4 Consequently, if $z(\lambda)$ is one of the curves in $\Lambda(\ell)$ moving away from the critical point ℓ, then

$$\frac{\partial \Re g(z(\lambda))}{\partial\lambda} \leq -\cos(\delta)\left|\frac{\partial z}{\partial\lambda}\right| \cdot |dg|.$$

3.2.5 Let ν denote a number bigger than the highest degree of zero of dg (which is finite since dg is pulled back from a compact Riemann surface). In the local constructions given above, $m \leq \nu$. We get an estimate $\Re g(\ell) - \Re g(z(\lambda)) \geq Cd(\ell, z(\lambda))^{\nu+1}$.

3.3 *Properties of* f:

3.3.1 f is essentially a flow, up to reparametrization. In other words $f(z,0) = z$ and there is a function $u(s,t)$ with $u(0,0) = 0$, $u(s,t) \geq t$, such that $f(f(z,t),s) = f(z,u(s,t))$ if $t+s \leq 1$. In fact, $f(z,t)$ is a flow in the region $t \leq \tau(z)$. In the region $t \geq \tau(z)$, $f(z,t)$ moves monotonically along the one dimensional set Λ.

3.3.2 When composed with g, the flow goes approximately in the negative real direction, in other words

$$\frac{\partial g(f(z,t))}{\partial t} \in S(\pm\delta) = \{re^{i\theta} : \theta \in [\pi - \delta, \pi + \delta]\}.$$

3.3.3 Consequently,

$$\frac{\partial(\Re g \cdot f)}{\partial t} \leq -\cos(\delta)|\frac{\partial f}{\partial t}| \cdot |dg|.$$

3.3.4 $f(z,t)$ is piecewise polynomial. In fact, it is piecewise linear in the region $t \leq \tau(z)$ and piecewise quadratic in the region $t \geq \tau(z)$.

3.3.5 The flow decreases $\Re g$ by a specified amount L, unless Λ is encountered first. By choosing appropriate sizes for the vectors V_i, we can arrange so that if $d(z,P) \leq R$, then either $\Re g(f(z,1)) \leq \Re g(z) - L$, or else $\tau(z) \leq 1/2$. In the latter case there is a critical point ℓ such that $f(z,t) \in \Lambda(\ell)$ for $t \geq \tau(z)$.

3.3.6 Within $\Lambda(\ell)$, the flow moves away from the critical points. It does so to an extent governed by L and σ. We can choose the constant C_1 in the flows f_Λ such that if $\tau(z) \leq 1/2$ and $f(z,\tau(z)) \in \Lambda(\ell)$, then

$$\Re g f(z,\tau(z)) - \Re g f(z,1) \geq \frac{L}{\sigma}(\Re g(\ell) - \Re g f(z,\tau(z))).$$

With these choices made, we obtain some bounds.

3.3.7 There is a number R' such that if $d(z,P) \leq R$ and $t \in [0,1]$ then $d(f(z,t),P) \leq R'$. In this region the gradient (in terms of both arguments z and t) is bounded by $|\nabla f| \leq C$ for some constant C.

3.3.8 There are constants C and $\varepsilon > 0$ such that: for any $x \in Z$ with $d(x,P) \leq R$, $t \in [0,1]$, and any $y > 0$, the one dimensional measure in of the set of τ such that $\Re g_{ij}(f_{ij}(z,t)) - \Re g_{ij}(f_{ij}(z,t+\tau)) \leq y$, is less than Cy^ϵ. This can be seen from the local pictures at the critical points.

In the situation of this paper, we will always let $d(x,y)$ be the complete metric on Z obtained by pulling back a smooth metric from the compact Riemann surface S. We have several functions g_{ij} on Z, whose differentials are pulled back from S. The singular metric $\inf_{i\neq j}|dg_{ij}|$ is pulled back from S, so it is complete and essentially comparable to the smooth metric.

In any given situation below, we will rotate so that the desired direction of flow is the negative real direction. Refer to the rotated functions g_{ij}, and construct Λ_{ij}, f_{ij} as described above. Recall that we will fix all of the flows f_{ii} to be equal to a given flow f^0, fixed regardless of rotations. Make the choice of f^0 generic with respect to all possible graphs $\Upsilon = \bigcup_{i\neq j}\Lambda_{ij}$ and points $v = \ell$, critical points of g_{ij}, for $i \neq j$.

We will often abreviate notation in one of the following ways. If $\ell = (\ell_1 \times \ldots \ell_n)$ is a critical point for g on Z_I, we will write $\Lambda(\ell)$ for $\Lambda_{i_0 i_1}(\ell_1) \times \cdots \times \Lambda_{i_{n-1} i_n}(\ell_n)$. And in this situation, where the indices are understood, we will drop the subscripts and refer to $\Lambda_{i_{k-1} i_k}(\ell_k)$ simply as $\Lambda(\ell_k)$.

4. Moving relative homology chains

In this section we will describe a formalism for moving relative homology chains. We will form a double complex to calculate relative homology, and then consider homotopies in this complex. It will be done explicitly, so as to facilitate getting bounds.

Z is a complex manifold of dimension one, the universal cover of the original Riemann surface S. We consider *indices* $I = (i_0, \ldots, i_n)$, saying $|I| = n$. For each such index let Z_I be the space Z^n. Let

$$Z_n = \coprod_{|I|=n} Z_I, \qquad Z_* = \coprod_I Z_I.$$

We will work with chains which are combinations of singular and de Rham chains. Our manifolds will have linear structures, in other words embeddings as open sets in vector spaces. By a *k-chain* on such a manifold Y we will mean a linear functional on the space of C^∞ differential k-forms on Y which can be expressed as a sum of components of the following form $h(u * H)$. Here H is a $k+l$ dimensional space, compact, with linear structure and algebraic boundary, together with $h : H \to Y$ a smooth algebraic map (in other words the map is given by coordinate functions which are algebraic over the ring of polynomial functions on H). It is contracted with a smooth differential l-form u on H. Such a chain provides a linear functional on the space of k-forms a by the rule

$$\langle h(u * H), a\rangle = \int_H u \wedge h^*(a).$$

The reader may think primarily of singular chains (corresponding to the case when u is just the function 1). The more general singular-de Rham chains arise because we use cutoff functions later in the argument. Still, we usually denote $\langle \eta, a\rangle$ by $\int_\eta a$.

These algebraic singular-de Rham chains are functorial with respect to continuous piecewise polynomial maps (even though more general types of currents are not). Suppose $f : Y \to Y'$ is continuous and piecewise polynomial, and suppose $h(u * H)$ is a k-chain on Y. The composition $fh : H \to Y'$ is continuous and piecewise polynomial. We may further subdivide H into finitely many pieces H_i (with algebraic boundaries) such that on each H_i, fh is polynomial. Let u_i be the restriction of u to H_i. Then define

$$f(h(u * H)) = \sum (fh)(u_i * H_i).$$

The boundary of $h(u * H)$ is equal to $h(du * H) \pm h(u * \partial H)$, which again has the same form—the boundary components of H are algebraic.

Define the following groups of chains: $C^k(Z_I)$ is the group of k-chains on Z_I, and $C^{k,n}(Z_*)$ is the group of k-chains on Z_n. Define a *pro-chain* η_* to be a collection of chains η_I on Z_I respectively, so that the group of pro-chains is $C^{*,*}(Z_*) = \prod_{k,n} C^{k,n}(Z_*)$. If η_* is a pro-chain then denote by η_I the part supported on Z_I, and let η_n denote the chain $\sum_{|I|=n} \eta_I$ on Z_n. Occasionally we will drop the asterisk subscript for brevity.

Define the *support* of a chain η to be the smallest closed subset $Supp_{Z_*}(\eta)$ of Z_* such that if a is a form which vanishes on the subset, then $\langle \eta, a \rangle = 0$. The support of a chain could be smaller than the union of the images of the polyhedra involved, because some cancellation might occur.

Recall that two points P and Q are fixed in Z. Make the convention that if $z = (z_1, \ldots, z_n)$ is a point in Z_I, with $|I| = n$, then z_0 denotes P and z_{n+1} denotes Q.

Each space Z_I is endowed with a holomorphic function g_I defined by

$$g(z_1, \ldots, z_n) = -g_{i_0}(z_0) + g_{i_0 i_1}(z_1) + \ldots + g_{i_{n-1} i_n}(z_n) + g_{i_n}(z_{n+1})$$

(recall that $g_{ij}(z) = g_i(z) - g_j(z)$). It is sometimes useful to think of Z_I or the union Z_* as a space relative to $\mathbf{C}$, using the function g. The function g gives rise to another version of the *support* of a chain η, namely

$$Supp_{\mathbf{C}}(\eta) = g(Supp_{Z_*}(\eta)) \subset \mathbf{C}.$$

There is also an integrand b_I assigned to each component Z_I. If $|I| = n$, then b_I is a holomorphic matrix-valued n-form on Z_I, given by

$$b_I(z_1, \ldots, z_n) = \mathbf{e}_{i_n i_0} b_{i_n i_{n-1}}(z_n) \cdots b_{i_1 i_0}(z_1).$$

We can consider g and b respectively as a function and a matrix valued form of top degree on Z_*.

There is a pro-chain β_* determined by the choice of path γ from P to Q:

$$\beta_I = \{(\gamma(t_1), \ldots, \gamma(t_n)) : 0 \le t_1 \le \ldots \le t_n \le 1\}.$$

We may assume that the path γ is linear with respect to the linear structure of Z, so β_* is actually piecewise linear. The formula for the monodromy is

$$m(t) = \int_{\beta_*} b e^{tg}.$$

The formula for the Laplace transform is

$$f(\zeta) = \int_{\beta_*} \frac{b}{g-\zeta}.$$

These formulas imply sums over the components corresponding to the different indices I.

Lemma 4.1 *The integral $f(\zeta)$ is well defined and convergent for large values of $|\zeta|$.*

Proof: Recall that

$$g_I(\gamma(t_1), \ldots, \gamma(t_n)) = \sum_{j=0}^{n} \int_{t_j}^{t_{j+1}} \gamma^* a_{i_j},$$

with the conventions $t_0 = 0$ and $t_{n+1} = 1$. Therefore

$$|g_I(\gamma(t_1), \ldots, \gamma(t_n))| \leq \int_0^1 \sup_i |\gamma^* a_i|.$$

This bound for $|g_I(z)|$, $z \in Supp_{Z_I}(\beta_I)$, is independent of I. Therefore $Supp_{\mathbf{C}}(\beta_*)$ is contained in a compact subset of $\mathbf{C}$. If $d(\zeta, Supp_{\mathbf{C}}(\beta_*)) \geq 1$ then $|(g-\zeta)^{-1}| \leq 1$ on the support of β_*. If $|I| = n$ then the size of β_I is $C^n/n!$. The number of indices I with $|I| = n$ is bounded by C^n, and the size of $b_I(z_1, \ldots, z_n)$ is bounded by C^n. This proves that the sum

$$f(\zeta) = \sum_I \int_{\beta_I} \frac{b_I}{g_I - \zeta}$$

converges.

This lemma shows that the infinite sum of Laplace transforms $f(\zeta) = \sum_I f_I(\zeta)$ satisfies condition (2.5.0). We will, in the rest of the paper, show that the sum satisfies the other conditions (2.5.1)-(2.5.5). Mainly this involves analytically continuing $f(\zeta)$ to other values of ζ by moving the cycle of integration β_* away from the locus of poles $g^{-1}(\zeta)$, keeping the same relative homology class. This process should preserve the bound which allows us to sum over $n = |I|$, so we need a formalism for moving chains in relative homology while keeping track of the size.

There are some canonical inclusions between the spaces Z_I which preserve the function g. These correspond to the tetrahedra on the boundaries of a tetrahedron β_I. An *elementary arrow* $\alpha : I' \to I$ is a choice of number l such that $i'_k = i_k$ for $k < l$ and $i'_k = i_{k+1}$ for $k \geq l$. In particular note that if $|I| = n$ then $|I'| = n-1$. We allow l to be in the range $0 \leq l \leq n$. If α is an elementary arrow, we obtain a map in the opposite direction between intervals of integers:

$$\alpha^+ : [0, n+1] \to [0, n]$$

defined by $\alpha^+(k) = k$ if $k \leq l$ and $\alpha^+(k) = k - 1$ if $k > l$. Thus α^+ is order preserving and maps l and $l+1$ to l but is otherwise one-to-one, and $\alpha(0) = 0$ and $\alpha(n+1) = n$. For each elementary arrow α we get a map $\alpha : Z_{I'} \to Z_I$ defined by $\alpha(z)_k = z_{\alpha^+(k)}$. Thus $\alpha(z)_k = z_k$ if $k \leq l$ and $\alpha(z)_k = z_{k-1}$ if $k > l$. In particular $\alpha(z)_l = \alpha(z)_{l+1}$. Recall that $z_0 = P$ and $z_n = Q$ by convention ($z \in Z_{I'}$ with $|I'| = n - 1$). Notice that $g(\alpha(z)) = g(z)$. Define $sgn(\alpha) = (-1)^{l+n}$ (or whatever is compatible with the sign conventions for the boundary operator ∂).

The images $\alpha Z_{I'}$ in Z_I form a geometric simplex like the T^{n-1} described in §1. Their intersections are the images $\alpha Z_{I''}$ in $Z_{I'}$, and so forth. With this as motivation, define an operator A on pro-chains as follows. If η is a pro-chain, then $A\eta$ is the pro-chain defined by

$$(A\eta)_I = \sum_{\alpha : J \to I} sgn(\alpha)\alpha(\eta_J).$$

Lemma 4.2 *$A^2 = 0$ and $\partial\beta = A\beta$.*

Proof: The proof of the first formula is the same as the proof that $\partial^2 = 0$ in singular homology. For the second formula, note that if $|I| = n$ then $\partial\beta_I$ is a sum of tetrahedra of dimension $n - 1$ given by the $(t_1, \ldots, t_n)$ where two adjacent t_i and t_{i+1} are equal. These tetrahedra are the $\alpha(\beta_{I'})$ for $\alpha : I' \to I$. One has to check that the signs are right (which is easier said than done).

The operators ∂ and A are the differentials of a double complex $C^{k,n}(Z_*)$. The degrees of ∂ and A are $(-1, 0)$ and $(0, 1)$ respectively. Note that $\partial A = A\partial$ (since the operator A was obtained by pushing forward chains using continuous maps). The total differential is $\partial - (-1)^{k-n}A$.

Lemma 4.3 *If κ, η and η' are pro-chains such that*

$$\eta = \eta' + (\partial + A)\kappa$$

and such that the supports $Supp_{\mathbf{C}}(\kappa)$, $Supp_{\mathbf{C}}(\eta)$, and $Supp_{\mathbf{C}}(\eta')$ do not meet $\zeta \in \mathbf{C}$, then

$$\int_{\eta} \frac{b}{g-\zeta} = \int_{\eta'} \frac{b}{g-\zeta}$$

(assuming the two integrals are convergent).

Proof: For each index I, we have

$$\eta_I = \eta'_I + \partial\kappa_I + \sum_{\alpha:J\to I} sgn(\alpha)\alpha\kappa_J.$$

The integrand $b/(g-\zeta)$ is a holomorphic form of top degree, so its exterior derivative is zero. Since ζ is not in $Supp_{\mathbf{C}}(\kappa_I)$, this implies

$$\int_{\partial\kappa_I} \frac{b}{g-\zeta} = 0.$$

Similarly, the restriction of a holomorphic form of top degree to an analytic subvariety of smaller dimension is equal to zero, so for any elementary arrow $\alpha : J \to I$,

$$\int_{\alpha\kappa_J} \frac{b}{g-\zeta} = 0.$$

Therefore

$$\int_{\eta_I} \frac{b}{g-\zeta} = \int_{\eta'_I} \frac{b}{g-\zeta}.$$

If the sums over I converge, then the resulting integrals over pro-chains η and η' will be equal.

With this lemma in mind, our goal is to replace β by a series of pro-chains η, each differing from the last by a chain of the form $(\partial + A)\kappa$, in such a way that the domain of analyticity in ζ is extended.

We do this by applying the flows constructed in the previous section, coordinate by coordinate. There are some complications for the problem of moving cycles in relative homology. After completing the formalism, we will describe the geometric pictures in some low cases.

For technical reasons we need a buffer into which a request for flowing can be placed, to be enacted later. We accomplish this by introducing some new coordinates, representing the requested flows. That way if the same flow is requested several times, it is only used for unit time (instead of several unit times). Define a space $X = Z \times [0,1]$. Define spaces X_I in the same manner as above. This gives the union X_* and the spaces of chains $C^{k,n}(X_*)$. Let $p : X \to Z$ or $p : X_I \to Z_I$ be the projection. Consider Z_I as a subspace of X_I by setting the second coordinates to zero.

We have a function

$$G : X_I \times [0,1]^n \to X_I$$

defined as follows.

$$G(z_1, \ldots, z_n, s_1, \ldots, s_n, t_1, \ldots, t_n) = (z_1, \ldots, z_n, r_1, \ldots, r_n)$$

with $r_i = min(s_i + t_i, 1)$. Define a map $H : X_I \to X_I$ by

$$H(z_1, \ldots, z_n, t_1, \ldots, t_n) = G(z_1, \ldots, z_n, 1, \ldots, 1).$$

If η is a chain on X_I, set

$$K_k \eta = G(\eta, 1, \ldots, 1, [0,1], 0, \ldots, 0)$$

with the interval $[0,1]$ occurring in the kth place. Then set

$$K\eta = \sum_{k=1}^{n} K_k \eta.$$

The orientation of $K_k\eta$ is to be chosen so that

$$\partial K\eta + K\partial\eta = H\eta - \eta.$$

K is a chain homotopy from the identity to H.

Now we use the flows which have been constructed in the previous section. For each i, j there is a function $f_{ij}(z,t) : X = Z \times [0,T] \to Z$. For all i, f_{ii} is our given fixed flow f^0. Using these functions in each factor we get a function

$$F : X_I \to Z_I$$

defined by

$$F(z_1, \ldots, z_n, t_1, \ldots, t_n) = (y_1, \ldots, y_n)$$

with $y_k = f_{i_{k-1}i_k}(z_k, t_k)$. This function can be thought of as applying the relevant flows and then resetting all second coordinates to zero. We also have to apply flows and reset the appropriate second coordinates to zero, when the maps α are used. Thus if $\alpha : I' \to I$ is an elementary arrow, define a map $\alpha : X_I \to X_I$ as follows. Let l be the number which defines α, so $\alpha^+(l) = \alpha^+(l+1) = l$. Set

$$\alpha(z_1, \ldots, z_{n-1}, t_1, \ldots, t_{n-1}) =$$

$$(z_1, \ldots, z_{l-1}, y, y, z_{l+1}, \ldots, z_{n-1}, t_1, \ldots, t_{l-1}, 0, 0, t_{l+1}, \ldots, t_{n-1})$$

with $y = f_{i_{l-1}i_l}(z_l, t_l)$. If $l = 0$ then

$$\alpha(z_1, \ldots, z_{n-1}, t_1, \ldots, t_{n-1}) = (P, z_1, \ldots, z_{n-1}, 0, t_1, \ldots, t_{n-1})$$

and if $l = n$ then Q is similarly inserted at the right.

Define an operation A on pro-chains on $\coprod X_I$ the same as before, in other words

$$(A\eta)_I = \sum_{\alpha : I' \to I} sgn(\alpha)\alpha(\eta_I).$$

Note that A and F commute.

The map H and the homotopy K do not preserve the subspaces $\alpha X_{I'} \subset X_I$. Therefore they do not send relative homology chains to relative homology chains. Thus we must define some more operations. Suppose η is a pro-chain in $\sum C^{n,n}(Z_*) \subset \sum C^{n,n}(X_*)$. Set

$$\nu = (\partial - A)\eta.$$

Note that $(\partial + A)\nu = 0$ by Lemma 4.2. Define a chain $\varphi(\eta)$ by the equation

$$\varphi = \eta + AK\varphi.$$

This makes sense because AK increases dimension by one, so

$$(1 - AK)^{-1} = 1 + AK + AKAK + \ldots$$

is well defined on pro-chains. Similarly define $\tau(\eta)$ by

$$\tau = H\varphi - KA\tau$$

and $\psi(\eta)$ by

$$\psi = K\nu - KA\psi.$$

Note that

$$(1 + KA)(\tau - \psi) = H\varphi - K\nu.$$

Lemma 4.4 *$\partial\varphi = A(\tau - \psi) + \nu$ and $\tau = \psi + \varphi + \partial K\varphi$. Hence $\tau = \eta + \psi + (\partial + A)K\varphi$.*

Proof: We have

$$\begin{aligned}\partial\varphi &= \partial\eta + \partial AK\varphi \\ &= A\eta + \nu + A\partial K\varphi \\ &= A\eta + \nu + A(H\varphi - \varphi - K\partial\varphi).\end{aligned}$$

But $A\varphi = A\eta + AAK\varphi = A\eta$ by Lemma 4.2, so

$$\partial\varphi = \nu + AH\varphi - AK\partial\varphi$$

which can be rewritten

$$(1 + AK)\partial\varphi = AH\varphi + \nu.$$

On the other hand

$$\begin{aligned}(1 + AK)(A(\tau - \psi) + \nu) &= A(1 + KA)(\tau - \psi) + \nu + AK\nu \\ &= A(H\varphi - K\nu) + \nu + AK\nu \\ &= AH\varphi + \nu\end{aligned}$$

so by comparing, $\partial\varphi = A(\tau - \psi) + \nu$. Now

$$\begin{aligned}\partial K\varphi &= -K\partial\varphi + H\varphi - \varphi \\ &= -KA(\tau - \psi) - K\nu + H\varphi - \varphi \\ &= (H\varphi - KA\tau) - (K\nu - KA\psi) - \varphi \\ &= \tau - \psi - \varphi\end{aligned}$$

as required. The last statement follows immediately from the equation for φ.

Examples

We will discuss in detail the cases $|I| = 0, 1, 2$. These should demonstrate the motivation for the formalism. Assume for simplicity that $\eta_* = \beta_*$ is the pro-chain determined by a path γ. In particular, $\nu = (\partial - A)\eta$ will be zero, so ψ will be zero and we can ignore those terms.

The case $|I| = 0$ is somewhat exceptional. The index is $I = (i)$; $Z_{(i)}$ is a single point (as is $X_{(i)}$); and the value of $g_{(i)}$ at this point is equal to $g_i(Q)$. The single point should be considered as a critical point for g. The chain $\eta_{(i)} = \beta_{(i)}$ is zero dimensional, the point simply taken with multiplicity one. There are no elementary arrows with target (i). The new chains $\varphi_{(i)}$ and $\tau_{(i)}$ are equal to $\eta_{(i)}$. In fact, throughout all applications of the procedures, the chain $\eta_{(i)}$ will remain the same. The integral $m_{(i)}(t)$ is equal to the elementary matrix $\mathbf{e}_{ii}$ times $e^{tg_i(Q)}$. The sum over i gives the initial term in the expansion, corresponding to the identity matrix in the iteration expansion for $E(Q,t)m(Q,t)$.

When $|I| = 1$ we have $I = (i,j)$. The space $Z_{(i,j)}$ is just equal to Z. There are two elementary arrows

$$\alpha_0 : (j) \to (i,j), \qquad \alpha_1 : (i) \to (i,j).$$

The images of the corresponding inclusions are $\alpha_0(Z_{(j)}) = P \in Z_{(i,j)} = Z$ and $\alpha_1(Z_{(i)}) = Q \in Z_{(i,j)} = Z$. The function is

$$\begin{aligned} g_{(i,j)}(z) &= g_{ij}(z) + g_j(Q) \\ &= g_i(z) - g_j(z) + g_j(Q) \\ &= \int_P^z a_i + \int_z^Q a_j. \end{aligned}$$

Note that this agrees with $g_{(j)}$ on P and $g_{(i)}$ on Q. The chain $\eta_{(i,j)}$ is the path from P to Q given by γ. Note that

$$\partial\eta_{(i,j)} = Q - P = \alpha_1(\eta_{(i)}) - \alpha_0(\eta_{(j)}) = (A\eta)_{(i,j)}.$$

Apply the moving process, noting that $K\varphi_{(i)} = K\varphi_{(j)} = 0$ so $\varphi_{(i,j)} = \eta_{(i,j)}$.

$$\begin{aligned} F\tau_{(i,j)} &= FH\varphi_{(i,j)} - FK(A\tau)_{(i,j)} \\ &= FH\eta_{(i,j)} - FKP + FKQ. \end{aligned}$$

Thus $\eta_{(i,j)}$ becomes homologous to a sum of three pieces,

$$\eta_{(i,j)} + (\partial + A)FK\varphi = F\tau_{(i,j)} = FH\eta_{(i,j)} + FKQ - FKP.$$

Note that $FH\eta_{(i,j)} = f_{ij}(\eta_{(i,j)}, 1)$, while $FKQ = f_{ij}(Q,[0,1])$ and $FKP = f_{ij}(P,[0,1])$. The picture is

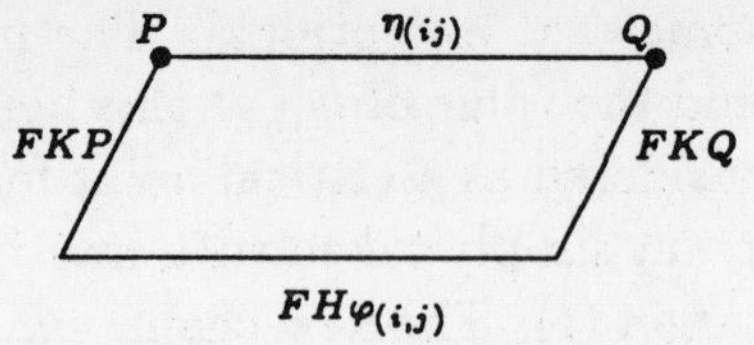

Suppose $|I| = 2$. This means that $I = (i, j, k)$. The space is $Z_{(i,j,k)} = Z \times Z$. There are three elementary arrows

$$\begin{aligned}\alpha_0 : (j, k) &\rightarrow (i, j, k),\\ \alpha_1 : (i, k) &\rightarrow (i, j, k),\\ \alpha_2 : (i, j) &\rightarrow (i, j, k).\end{aligned}$$

The images of the resulting maps are $\alpha_0(Z_{(j,k)}) = P \times Z \subset Z \times Z$ and similarly $\alpha_2(Z_{(i,j)}) = Z \times Q$, while $\alpha_1(Z_{(i,k)})$ is the diagonally embedded $Z \subset Z \times Z$. The chain $\eta_{(i,j,k)} = \beta_{(i,j,k)}$ is the triangle consisting of points $(z, w) = (\gamma(t_1), \gamma(t_2))$ for $0 \leq t_1 \leq t_2 \leq 1$. The boundary is contained in the geometric triangle formed by the images of the elementary arrows,

$$\partial\eta_{(i,j,k)} = \alpha_0(\eta_{(j,k)}) - \alpha_1(\eta_{(i,k)}) + \alpha_2(\eta_{(i,j)}) = (A\eta)_{(i,j,k)}.$$

The function is

$$g_{(i,j,k)}(z, w) = g_i(z) - g_j(z) + g_j(w) - g_k(w) + g_k(Q).$$

This agrees with the functions $g_{(j,k)}$, $g_{(i,k)}$, and $g_{(i,j)}$, on $P \times Z$, the diagonal Z, and $Z \times Q$ respectively. The chain $F\varphi_{(i,j,k)}$ is obtained by adding to $\eta_{(i,j,k)}$ the homotopies used in the case $|I| = 1$ on the geometric boundary pieces, namely $\alpha_0 FK\eta_{(j,k)}$, $\alpha_1 FK\eta_{(i,k)}$, and $\alpha_2 FK\eta_{(i,j)}$. The picture is

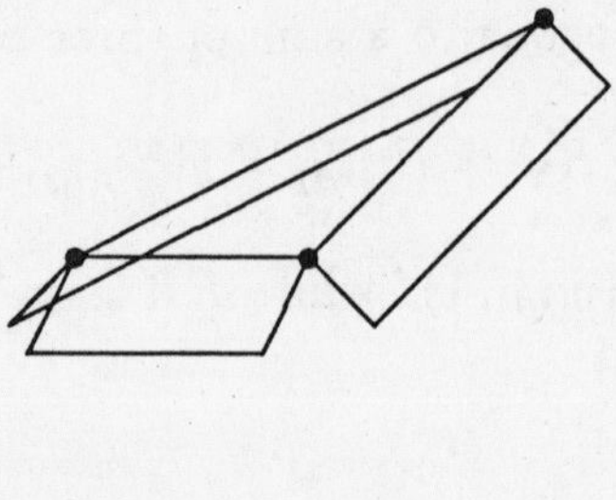

Note that

$$\partial F\varphi_{(i,j,k)} = \alpha_0 F\tau_{(j,k)} - \alpha_1 F\tau_{(i,k)} + \alpha_2 F\tau_{(i,j)}.$$

This is a sum of nine pieces, because each $F\tau$ from the case $|I| = 1$ is a sum of three pieces.

When we move $\varphi_{(i,j,k)}$ by the homotopy $FK\varphi_{(i,j,k)}$, it becomes homologous to $F\tau_{(i,j,k)}$. The new chain $F\tau_{(i,j,k)}$ is a sum of pieces of two types. The result of applying the homotopy to the boundary of $\varphi_{(i,j,k)}$ is $(FKA\tau)_{(i,j,k)}$. This is a homotopy contained in, though not quite equal to

$$(FKFA\tau)_{(i,j,k)} = FK\alpha_0 F\tau_{(j,k)} - FK\alpha_1 F\tau_{(i,k)} + FK\alpha_2 F\tau_{(i,j)}$$

(The use of buffer spaces X_I causes some complications, although it actually serves to reduce the sizes of things). There are nine pieces here.

The other type of piece in $F\tau_{(i,j,k)}$ comes from $FH\varphi_{(i,j,k)}$. This is the result of applying the flow to $\varphi_{(i,j,k)}$ to time 1. Again, because of our use of the buffers, it is not quite the same as $FHF\varphi_{(i,j,k)}$, but the latter provides an adequate picture. There are four pieces, one for each piece of $\varphi_{(i,j,k)}$. Thus the chain $F\tau_{(i,j,k)}$ is a sum of thirteen pieces altogether.

From this näive description of the procedure, it seems that the number of pieces in these chains will grow rapidly. In the rough count started above (ignoring subtleties involving the buffers X_I), set T_n equal to the number of pieces in $F\tau_I$ for $|I| = n$, and P_n equal to the number of pieces in $F\varphi_I$ for $|I| = n$. We have $T_0 = 1$ and $P_1 = 1$, and

$$T_n \leq P_n + (n+1)T_{n-1}, \quad P_n \leq 1 + (n+1)P_{n-1}.$$

The inequality is improved even in the case $|I| = 1$ above, as $P_1 = 1 < 3$, incorporating the first degeneracy $K\varphi_{(i)} = K\varphi_{(j)} = 0$.

From this crude counting, the number of pieces seems to grow at a rate worse than $n!$, because of the proliferation of boundary terms. Luckily, most of these pieces become degenerate if the procedure is chosen carefully. In order to get this degeneracy, it is necessary to apply the flows coordinate by coordinate as we have done, in order relative to the natural ordering of the factors in Z_I. The degeneracy provides a bound of C^n on the sizes of the chains, as will be seen in the next section.

Supports

Here are the support properties of the chains we have constructed above. Recall that $S(x, \pm\delta) \subset \mathbf{C}$ denotes the sector

$$S(x, \pm\delta) = \{x - re^{i\theta} : r \geq 0, -\delta < \theta < \delta\}.$$

If U is a subset of $\mathbf{C}$ then $S(U, \pm\delta)$ denotes the union

$$S(U, \pm\delta) = \bigcup_{x \in U} S(x, \pm\delta).$$

Assume that the flows f_{ij} have been chosen as described in §3, with respect to a number L, an angular error δ, and a number σ. Construct pro-chains on X_*

$$\varphi = \varphi(\eta), \quad \psi = \psi(\eta), \quad \tau = \tau(\eta).$$

Consider their images $F\varphi$, $F\psi$, $F\tau$ as well as the pro-chain $FK\varphi$ in Z_*. The following lemma shows that the process described above moves the supports in $\mathbf{C}$ in approximately the negative real direction, to within angular error δ.

Lemma 4.5

$$Supp_{\mathbf{C}}(F\varphi) \subset S(Supp_{\mathbf{C}}(\eta), \pm\delta)$$

$$Supp_{\mathbf{C}}(FK\varphi) \subset S(Supp_{\mathbf{C}}(\eta), \pm\delta)$$

$$Supp_{\mathbf{C}}(F\tau) \subset S(Supp_{\mathbf{C}}(\eta), \pm\delta)$$

and

$$Supp_{\mathbf{C}}(F\psi) \subset S(Supp_{\mathbf{C}}((\partial - A)\eta), \pm\delta).$$

Proof: The first contention is that for any chain η in X_*,

$$Supp_{\mathbf{C}}(FK\eta) \subset S(Supp_{\mathbf{C}}(F\eta), \pm\delta).$$

In the right side of this inclusion, the definition of support is being misinterpreted slightly—we wouldn't want any accidental cancellations to make the support of $F\eta$ too small; so the convention is that $Supp_{\mathbf{C}}(F\eta)$ should be taken to mean $g(F(Supp_{X_*}(\eta)))$, with similar conventions in the relevant places below. Here is why the contention is true. By the definition of K, the support of $FK\eta$ is contained in the support of $FG(\eta, [0,1], \ldots, [0,1])$. Condition (3.3.1) satisfied by the flows implies that this support is contained in the support obtained by first applying the operation F to η, then using the flows. Thus the support of $FK\eta$ is contained in the support of $F(F\eta, [0,1], \ldots, [0,1])$.

Now use condition (3.3.2) on the flows, coupled with the fact that the function $g_I(z_1, \ldots, z_n)$ is a sum of functions $g_{i_{k-1}i_k}(z_k)$. These imply that for $(z_1, \ldots, z_n) \in Z_I$,

$$Supp_{\mathbf{C}}(F(z_1, \ldots, z_n, [0,1], \ldots, [0,1])) \subset S(g_I(z_1, \ldots, z_n), \pm\delta).$$

Letting $(z_1, \ldots, z_n)$ range over the support of $F\eta$, this serves to prove the first contention.

If η is any chain, then $Supp_{\mathbf{C}}(FA\eta) = Supp_{\mathbf{C}}(F\eta)$. Inductively applying this fact and the first contention, it follows that for a chain η in Z_*,

$$Supp_{\mathbf{C}}(F(AK)^r\eta) \subset S(Supp_{\mathbf{C}}(\eta). \pm \delta)$$

By the formula for φ, this gives the first inclusion in the lemma. The second and third inclusions follow similarly. In the argument for τ, note that the first contention above would have worked just as well had $FK\eta$ been replaced by $FH\eta$. The fourth inclusion follows by the same reasoning applied to $\nu = (\partial - A)\eta$.

5. The main lemma

In this section we will count the cells in the chains φ, τ, and ψ that were defined in the previous section. Note that

$$\varphi = \sum_r (AK)^r \eta$$

$$\tau = \sum_r (-KA)^r H\varphi$$

$$\psi = \sum_r (-KA)^r K\nu$$

We will show that the number of nondegenerate cubical cells in one of these chains is bounded by C^n, by parametrizing the cells with trees.

Suppose z is a point in Z_I, with $|I| = n$. Consider the chain $F(KA)^r z$. It is a sum of cells of the form

$$FK_{k_r}\alpha_r \dots K_{k_1}\alpha_1 z.$$

Each of these cells is an r-cube. Our main construction will be to describe the points in these cells using graphs (which are trees).

Fix a sequence $\alpha_1, \dots, \alpha_r$. In particular there is a sequence of indices $I = I_0, I_1, \dots, I_r$ such that $\alpha_j : I_{j-1} \to I_j$. We will associate a graph to this choice as follows. The vertices are arrayed in $r+1$ rows, with the top row having $n+2$ vertices and bottom row having $n+r+2$ vertices. The jth row from the top has $n+j+2$ vertices. The vertices are numbered from right to left in each row, beginning with 0, and we denote the kth vertex in the jth row by v_{jk}. The vertices at the ends of the rows, v_{j0} and $v_{j(n+j+1)}$, are called *side* vertices. The edges of the graph go from vertices in one row to vertices in the next. There is an edge connecting $v_{j-1\,i}$ to v_{jk} if and only if $\alpha_j^+(k) = i$. Thus in each row except the bottom one, there is exactly one vertex with two edges emanating from below, and all of the other vertices have one edge below. The edges, when drawn as straight lines, do not intersect, because the maps α^+ are order preserving. The edges drawn from v_{j0} to $v_{j+1\,0}$ and from $v_{j(n+j+1)}$ to $v_{(j+1)(n+j+2)}$ are called *side* edges.

Suppose that we assign a number between 0 and 1 to each edge except the side edges. Then we obtain a point $u \in Z_{I_r}$ as follows. For each j, let $t_{j1}, \dots, t_{j\,n+j}$ denote the numbers assigned to the edges above the vertices $v_{j1}, \dots, v_{j\,n+j}$ respectively. Define $u_j \in X_{I_j}$ inductively for $j = 0, \dots, r$. First, $u_0 = z$. Then set $u_j = G(\alpha_j(u_{j-1}), t_{j1}, \dots, t_{j\,n+j})$. Set $u = F(u_r)$.

We can decompose the graph into *strands*, with the strands joining *forks*. The forks are the vertices which are connected to three edges (in other words the vertices v_{jk} such that $\alpha^+_{j+1}(k) = \alpha^+_{j+1}(k+1) = k$), as well as, by convention, the top and bottom vertices. The strands are the unbroken sequences of edges joining forks, in other words the sequences of edges which meet at interior vertices with only two edges. Side strands are those consisting of side edges. The graph formed by the forks and strands considered as vertices and edges, is a union of binary trees. If a number is assigned to each non-side edge, then one obtains a number for each non-side strand as follows. Suppose σ is a strand, composed of edges $e_1, \ldots, e_m$. Set

$$t(\sigma) = min(1, t(e_1) + \ldots + t(e_m)).$$

In the above construction, the point u depends only on the numbers $t(\sigma)$ assigned to the strands. Here is another description of the construction of u. For each strand σ there are indices $i(\sigma)$ and $j(\sigma)$, representing the indices corresponding to the left and right sides of the edges in the strand, respectively. If the strand σ contains an edge ending in a vertex v_{jk}, then $i(\sigma) = i_{j,k-1}$ and $j(\sigma) = i_{j,k}$. (The notation $i(e)$ and $j(e)$ will also be used for an edge e.) Realize the tree geometrically, with a strand σ represented by a line segment of length 1. Let T denote the geometric realization of the tree. Then the function t from the set of strands into $[0,1]$, and the initial point z, determine a map $\Psi_{z,t} : T \to Z$. Write $z = (z_1, \ldots, z_n)$. The top vertices of the tree go to the points $z_k \in Z$. The left and right side strands are mapped to P and Q respectively. If σ is any strand, $\Psi_{z,t}$ maps the segment corresponding to σ into Z using the flow $f_{i(\sigma)j(\sigma)}$, beginning with the point corresponding to the fork v at the top of σ, and moving at speed $t(\sigma)$. The beginning point $\Psi_{z,t}(v)$ has already been constructed inductively. If p is a point on the segment σ, at distance y below the fork v, $\Psi_{z,t}(p) = f_{i(\sigma)j(\sigma)}(\Psi_{z,t}(v), t(\sigma)y)$. Finally, the the values of $\Psi_{z,t}$ on the $n+r$ bottom vertices provide the points $u_1, \ldots, u_{n+r}$ to determine $u = u(z,t) \in Z_{I_r}$.

The above construction describes the points in the cell

$$FK_{k_r}\alpha_r \ldots K_{k_1}\alpha_1 z.$$

Let $s_1, \ldots, s_r$ denote parameters in $[0,1]$. Assign numbers to the non-side edges of the graph according to the choice of $k_1, \ldots, k_r$. In the jth row of edges, assign the number s_j to the edge above the vertex v_{jk_j}. Assign the number 1 to the edges above the vertices v_{ji} for $i < k_j$ and assign the number 0 to the edges above the vertices v_{ji} for $i > k_j$. By the construction described above we get points $u(s_1, \ldots, s_r)$. It is easy to see from the definitions in the previous section that the cell $FK_{k_r}\alpha_r \ldots K_{k_1}\alpha_1 z$ is an r-dimensional cube formed of the points $u(s_1, \ldots, s_r)$. Furthermore, because of the fact that the points $u(s_1, \ldots, s_r)$ only depend on the numbers assigned to the strands, the cell is degenerate (that is, equivalent to zero because it is supported in a smaller dimension) if in any strand containing one edge marked s_j, there is another edge marked either 1 or s_j. In other words the cell is nondegenerate only if, in any strand containing an edge marked s_j, all of the other edges are marked 0.

Proposition 5.1 *There is a constant C, independent of n, z, or r, such that the number of nondegenerate cells as above in $F(KA)^r z$ is less than C^{n+r}.*

Proof: We will prove that there are less than C^{n+r} choices of $\alpha_1, \ldots, \alpha_r$, and $k_1, \ldots, k_r$, such that the condition for nondegeneracy described in the previous paragraph holds. As a matter of notation, let $l_1, \ldots, l_r$ be the numbers which determine $\alpha_1, \ldots, \alpha_r$ respectively. We now list the conditions for nondegeneracy, and then count the number of pairs of sequences $\{k_i\}$, $\{l_i\}$ which satisfy those conditions. For the purposes of this argument, consider the left side edges to have been assigned the number 1 and the right side edges to have been assigned the number 0.

First of all, if $l_j > k_{j-1}$ then the cell is degenerate. For then the picture is one of

$k_j < k_{j-1}$

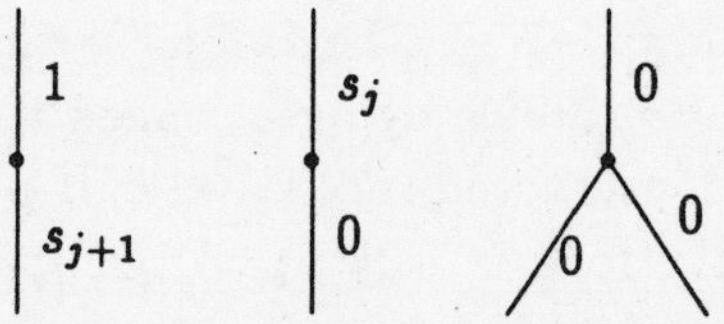

$k_j = k_{j-1}$

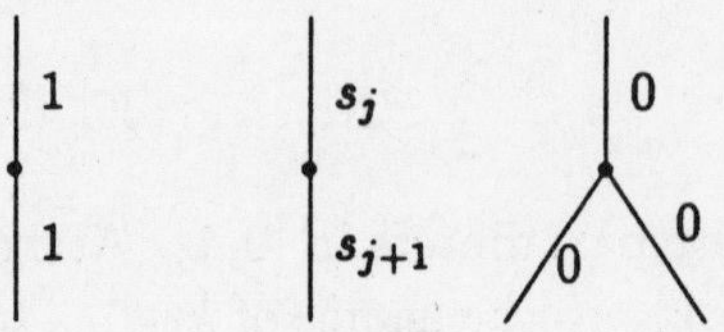

$k_j > k_{j-1}$

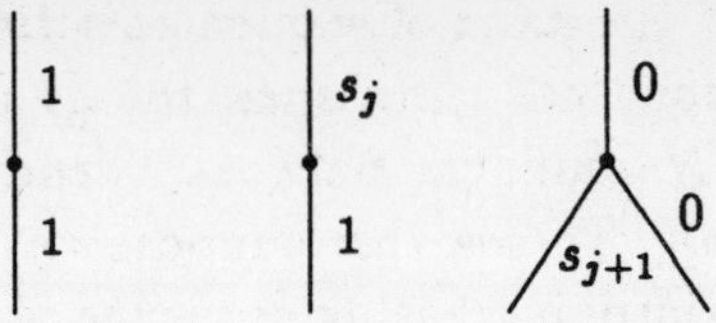

with the possibilities for the result of choosing $k_j < k_{j-1}$, $k_j = k_{j-1}$, or $k_j > k_{j-1}$, shown in order. None of these choices works. Therefore our first condition is $l_j \leq k_{j-1}$.

Given this first condition, it follows that all of the edges to the left of the l_jth one in the $j-1$st row are marked 1.

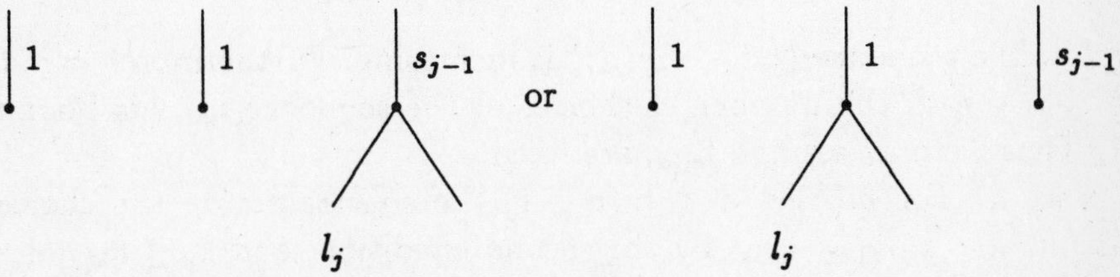

Therefore $k_j \geq l_j$, which is the second condition.

The third condition is as follows. Suppose that $l_j < k_{j-1}$. Then for any $i \geq j$ we have $k_i \leq l_j + 1 + i - j$. To see this consider for example the picture where the equality is attained:

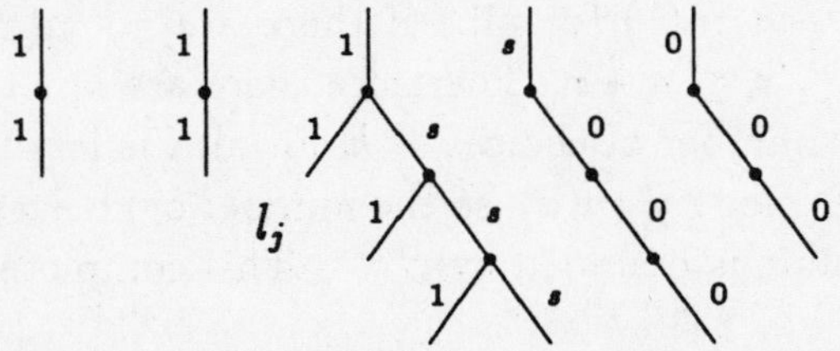

No edge marked s can be attached to the strands beginning at edges to the right of the l_jth edge in the $j-1$st row, unless one of those strands ends in a branch point. But inductively no branch point could be put there because of the first condition.

Now we count the pairs of sequences satisfying these three conditions. The principal fact about counting such things is that the number of increasing sequences of $\leq N$ numbers between 0 and N is bounded by C^N for some C ($C = 4$ works). To see this suppose r_i is the increasing sequence. Then $s_i = r_i + i$ is a strictly increasing sequence of numbers between 1 and $2N$. The number of sequences s_i is less than the number of subsets of $[1, 2N]$, which is 4^N.

Now turn to the sequences $\{k_i\}$ and $\{l_i\}$. Let $m(1), \ldots, m(q)$ denote the indices such that $l_m < k_{m-1}$. (Note that we may have $q = 0$.) The $m(i)$ are an increasing sequence between 1 and r, so there are less than C^r possibilities. Thus we may assume that the $m(i)$ are fixed.

If $i > m(a)$ then $k_{i-1} \leq l_{m(a)} + i - m(a)$. In particular $l_i \leq l_{m(a)} + i - m(a)$, so

$$l_{m(a+1)} - m(a+1) \leq l_{m(a)} - m(a).$$

Therefore the sequence $\{l_{m(a)} - m(a)\}$ is increasing. Furthermore $-r \leq l_{m(a)} - m(a) \leq n + r$ so the number of choices of the sequence $l_{m(a)}$ is bounded by C^{n+r}. Thus we may assume $l_{m(a)}$ are fixed.

Fix an a. For $m(a) \leq i < m(a+1)$ the constraints on the choice of k_i are as follows. $k_{m(a)} \geq l_{m(a)}$ by the second condition, and $k_i \leq l_{m(a)} + m(a+1) - m(a)$ by the third condition. Furthermore for $m(a) < i < m(a+1)$ we have $k_{i-1} = l_i \leq k_i$ (the equality is by the first condition and the definition of the $m(a)$ and the inequality is by the second condition). Therefore the k_i for $m(a) \leq i < m(a+1)$ are an increasing sequence of $m(a+1) - m(a)$ numbers in a range of length $m(a+1) - m(a)$. There are less than $C^{m(a+1)-m(a)}$ choices for this sequence. Thus the total number of choices of the k_i for all $i \geq m(1)$ is less than C^r. For $i < m(1)$ (or all i if there are no $m(a)$) we again have $k_{i-1} = l_i \leq k_i$, and $0 \leq k_i \leq n + r$. Therefore there are less than C^{n+r} choices there. Thus the total number of choices of k_i for all i is less than C^{n+r}. These choices determine the l_i for $i \neq m(a)$, so the number of choices of the sequences satisfying the constraints is bounded by C^{n+r}. This completes the proof of the proposition.

We now prove a proposition which will be used later, in §9. It roughly says that the marked strands in any tree which occurs above are equally distributed.

Proposition 5.2 *In any tree with marked strands which occurs above, there is a map from the forks (except the ones at the top) to the strands marked s, which is injective, and which maps each fork to an adjoining edge.*

Proof: Define the map M from forks to marked edges inductively by defining maps M_j on the part of the tree which comes from the operations $\alpha_1, \ldots, \alpha_j$ and $K_{k_1}, \ldots, K_{k_j}$. M_1 is trivial, for there are no forks except one at the top, at this stage. Suppose $j \geq 1$. When we apply α_{j+1} (corresponding to a number l_{j+1}) and then $K_{k_{j+1}}$, modify M_j to M_{j+1} as follows.

If $l_{j+1} < k_j$ then $k_{j+1} = l_{j+1}$ or $l_{j+1} + 1$, and in either case we can map the new fork corresponding to l_{j+1} to the new marked edge corresponding to k_{j+1}, which is right below it.

Now suppose $l_{j+1} = k_j$, and choose m so that $l_i = k_{i-1}$ for all $i > m$, but $l_m < k_{m-1}$. We claim that there is a row of consecutive strands marked s beginning at the fork corresponding to l_m and ending at the fork corresponding to l_{j+1}, such that all of the strands marked 0 immediately to the right of the fork l_{j+1} extend from forks in this row of edges, and such that under M_j each fork (except the last) in the row is mapped to the strand below it. We will construct the map M_{j+1} and also prove the claim inductively. If $k_{j+1} = l_{j+1}$ or $l_{j+1} + 1$ then the new marked edge is added below the new fork; we can add this to the row of edges, and let M_{j+1} be the same as M_j but also map the new fork to the new marked edge. If $l_{j+2} = k_{j+1}$ then the above claim will again hold. Now suppose $k_{j+1} > l_{j+1} + 1$. This means that the new marked edge is attached to one of the strands which was assigned 0, coming from a fork v in the row. Now obtain M_{j+1} as follows. For each fork strictly below v in the row of strands—including the last one, which is the new branch—let M_{j+1} map that fork to the marked edge above it in the row, instead of the one below as was the case with M_j. Let M_{j+1} map any fork above v to the edge below it in the row, as M_j did. Finally let M_{j+1} map the fork v to the new marked strand below it. Choose a new row of strands which consists of the old row above the fork v, and then the new marked strand below v instead of the old row below v. If $l_{j+2} = k_{j+1}$ then the vertex corresponding to l_{j+2} occurs at the end of the new row of edges, and the above claim will again hold. This completes the proof of the proposition.

6. Finiteness lemmas

In this section we will prove some statements which give finiteness conditions on the topology of the situation. These statements are the basis for the fact that the sets of singularities of the $f_n(\zeta)$ are locally finite in a way which is uniform in n.

Define a distance function $D_I(z,z')$ on Z_I as follows. For each i,j and $z_k, z'_k \in Z$ define

$$D_{ij}(z_k, z'_k) = \inf_{\omega} \int_0^1 \left| \frac{d}{dt} g_{ij}(\omega(t)) \right| dt$$

with the infimum taken over paths $\omega : [0,1] \to Z$ joining z_k to z'_k. Note that $D_{ii}(z_k, z'_k) = 0$. Now if $I = (i_0, \ldots, i_n)$, and $z = (z_1, \ldots, z_n)$, and $z' = (z'_1, \ldots, z'_n)$ are in Z_I, define

$$D_I(z,z') = \sum_{k=1}^{n} D_{i_{k-1}i_k}(z_k, z'_k).$$

Recall that we have fixed a flow $f^0(z,t)$ to be used as f_{ii} for all i.

Suppose $z \in Z_I$ and $w \in Z_J$, with $|I| = n$ and $|J| = n+m$. We say that w is *beyond z at distance $\leq M$* if there is a sequence of indices $I = I^0, I^1, \ldots, I^m = J$ with elementary arrows $\alpha^j : I^{j-1} \to I^j$, and a sequence of points $z = z^0, z^1, \ldots, z^m = w$ with $z^j \in Z_{I^j}$, such that

$$\sum_{j=1}^{m} D(\alpha^j(z^{j-1}), z^j) \leq M,$$

and such that if $i^j_k = i^j_{k-1}$ then $z^j_k = f^0(\alpha_j(z^{j-1})_k, t^j_k)$ for some t^j_k.

Recall that we have fixed a path γ from P to Q, and that this leads to a pro-chain β_*, a sum of tetrahedra.

Proposition 6.1 *Fix M. There is a number R such that if z is a point on β_* and if $w = (w_1, \ldots, w_n)$ is point in Z_I such that w is beyond z at distance less than or equal to M, then $d(w_k, P) \leq R$.*

Proof: Say $z = (z_1, \ldots, z_{n-m}) \in \beta_I$, with $z_j = \gamma(s_j)$, $0 \leq s_1 \leq \ldots \leq s_m \leq 1$. The sequence $\alpha_1, \ldots, \alpha_m$ defines a graph as in §5, arranged in m rows of edges and $m+1$ rows of vertices, with $n-m$ top vertices and n bottom vertices. There is a map Ψ from this graph into Z defined in the following way (reminiscent of though not identical to the discussion in §5). Let $z = z^0, z^1, \ldots, z^m = w$ be the sequence of points given by the definition. The image of the vertex v_{kl} is the point $\Psi(v_{kl}) = z_l^k$. Suppose e is an edge going from vertex v to vertex v', and let $i(e)$ and $j(e)$ denote the left and right indices at e. Define

$$\delta(e) \stackrel{def}{=} D_{i(e)j(e)}(\Psi(v), \Psi(v')).$$

The image of $\Psi(e)$ is defined to be the arc (parametrized $\Psi(e,t)$, $t \in [0,1]$) going from $\Psi(v)$ to $\Psi(v')$ which realizes the infimum

$$\delta(e) = D_{i(e)j(e)}(\Psi(v), \Psi(v')) = \int_0^1 |\frac{d}{dt} g_{i(e)j(e)} \Psi(e,t)| dt.$$

If $i(e) = j(e)$ then we may assume that $\Psi(e,t) = f^0(\Psi(v), at)$ where a is chosen so that $\Psi(v') = f^0(\Psi(v), a)$.

We may choose a function $\phi_0(x)$ with the following properties. First of all, ϕ_0 is an exhaustion function, in other words the subset $\{x : \phi_0(x) \leq R\}$ is compact in Z. Next,

$$|d\phi_0| \leq \inf_{i \neq j} |dg_{ij}|.$$

And finally, ϕ_0 is decreased by the flow f^0, in other words $\phi_0(f^0(x,t)) \leq \phi_0(x)$. The reason we may assume that ϕ_0 is an exhaustion function is that Z is the universal cover of a compact Riemann surface S, and the one forms dg_{ij} are pulled back from S. Thus the singular metric $\inf_{i \neq j} |dg_{ij}|$ is complete, so the second condition provides no obstruction to making ϕ_0 grow at infinity.

Suppose e is an edge going from v' to v''. If $i(e) \neq j(e)$, then

$$\phi_0(\Psi v'') - \phi_0(\Psi v') \leq \delta(e)$$

since

$$|\frac{d}{dt} \phi_0(\Psi(e,t))| \leq |\frac{d}{dt} g_{i(e)j(e)}(\Psi(e,t))|$$

by the second condition on ϕ_0. If $i(e) = j(e)$ then $\Psi v'' = f^0(\Psi v', t)$ so

$$\phi_0(\Psi v'') - \phi_0(\Psi v') \leq 0 = \delta(e)$$

by the third condition on ϕ_0. Suppose w_k is one of the images of the bottom vertices. There is a point z_l, and a sequence of edges $e_1, \ldots, e_m$ going respectively between vertices $v^0, v^1, \ldots, v^m$ with $\Psi(v^0) = z_l$ and $\Psi(v^m) = w_k$. Then

$$\phi_0(w_k) - \phi_0(z_l) = \sum_{i=1^m} (\phi_0(\Psi v^i) - \phi_0(\Psi v^{i-1})) \leq \sum_{i=1}^{m} \delta(e_i).$$

Our hypothesis that the point w is beyond points z at distance $\leq M$ means that $\sum_e \delta(e) \leq M$. In particular, the partial sum $\sum_{i=1}^{m} \delta(e_i)$ is less than M, so

$$\phi_0(w_k) \leq \phi_0(z_l) + M.$$

On the other hand, the hypothesis that $(z_1, \ldots, z_{n-m})$ is a point of β_* means that z_l lies on the path γ. The path is fixed, so there is an upper bound $\phi_0(\gamma(t)) \leq C$. Thus

$$\phi_0(w_k) \leq C + M.$$

Since ϕ_0 is an exhaustion function, there is a constant R depending only on ϕ_0, C, and M, such that $d(w_k, P) \leq R$.

Proposition 6.2 *For any fixed M there is a number N_0 with the following property. Suppose z is a point on β_*, $I = (i_0, \ldots, i_n)$ is an index, and $\ell = (\ell_1, \ldots, \ell_n)$ is a critical point for g_I on Z_I such that ℓ is beyond z at distance less than or equal to M. For each k with $i_{k-1} \neq i_k$, let $r(k)$ be the smallest positive number such that $i_{k-r(k)-1} \neq i_{k-r(k)}$. Then there are at most N_0 numbers k such that $i_{k-1} \neq i_k$ and $\ell_{k-r(k)} \neq \ell_k$. In particular the set S_M of values of $g(\ell)$ at critical points which are beyond β_* at distance $\leq M$, is finite.*

Proof: Employ the same mechanism as in the previous proof. There is a map Ψ from a tree to Z such that the ends of the tree go to the critical points $\ell_1, \ldots, \ell_n$. By the previous proposition, $d(\ell_k, P) \leq R$. For brevity, we will denote also by ℓ_k the vertices in the tree which are mapped to the critical points.

For each distinct point $\ell \in Z$ with $d(\ell, P) \leq R$ such that ℓ is a critical point of some g_{ij} with $i \neq j$, define a function $\phi(\ell)(x)$ with the following properties. $\phi(\ell)(x) = 1$ if there is an s such that $f^0(x, s) = \ell$.

$$\frac{d}{dt}\phi(\ell)(f^0(x,t)) \leq 0$$

for all x. If $\ell \neq \ell'$ then the supports of $\phi(\ell)$ and $\phi(\ell')$ are disjoint (this can be achieved since, by hypothesis, the flow f^0 is chosen generically). The support of $d\phi(\ell)$ is disjoint from any critical point in $d(z, P) \leq R$ (including ℓ itself). There is a constant C such that if an edge e goes from v to v',

$$\phi(\ell)(\Psi v') - \phi(\ell)(\Psi v) \leq C D_{i(e)j(e)}(\Psi v, \Psi v').$$

This is true even if $i(e) = j(e)$, by the condition that f^0 decreases ϕ.

There is a natural partial ordering of the forks in the graph: the fork at the end of a strand is below the fork at the start of the strand, and the ordering is obtained by taking the transitive closure of this relation. If v and v' are forks such that v' is below v then there is a sequence of branches $v = v_0, v_1, \ldots, v_m = v'$ and a sequence of strands $s_1, \ldots, s_m$ such that v_i is the end of s_i and the start of s_{i+1}. Thus there is a sequence of points $\Psi(v_0), \Psi(v_1), \ldots, \Psi(v_m)$ and edges $\Psi(e_i)$ connecting $\Psi(v_{i-1})$ to $\Psi(v_i)$. Let (i_k, j_k) be the indices associated to the edge e_k. Define

$$\Delta(v, v') = \sum_{k=1}^{m} D_{i_k j_k}(\Psi v_{k-1}, \Psi v_k).$$

If $\phi(\ell)(\Psi v_0) = 0$ and $\phi(\ell)(\Psi v_m) = 1$ then $\Delta(v, v') \geq (1/C)$ by the previous paragraph (this is where we utilize the functions $\phi(\ell)$).

Now we define a number $\Delta_1(\ell_{k-r}, \ell_k)$ whenever ℓ_{k-r} and ℓ_k are two points from the bottom of the graph such that $i_{k-r-1} \neq i_{k-r}$ and $i_{k-1} \neq i_k$. Δ_1 measures the total distance between ℓ_{k-r} and ℓ_k along edges of the graph, using the distances defined above. If there is a fork v above both ℓ_{k-r} and ℓ_k then assume that v is the smallest such fork in the partial ordering, and define

$$\Delta_1(\ell_{k-r}, \ell_k) = \Delta(v, \ell_{k-r}) + \Delta(v, \ell_k).$$

If there is no fork above both ℓ_k and ℓ_{k-r} then there are vertices v and v' on the top of the graph such that ℓ_{k-r} is below v and ℓ_k is below v'. In this case define

$$\Delta_1(\ell_{k-r}, \ell_k) = \Delta(v, \ell_{k-r}) + \Delta(v', \ell_k).$$

It is easy to see that

$$\sum_e \delta(e) = \frac{1}{2} \sum_k{}' \Delta_1(\ell_{k-r(k)}, \ell_k).$$

Here the sum is taken over those k such that $i_{k-1} \neq i_k$. For each such k, $r(k)$ is chosen so that $k - r(k)$ is the previous one. Under our hypothesis that $(\ell_1, \ldots, \ell_n)$ is beyond z at distance less than M, we get

$$\sum_k{}' \Delta_1(\ell_{k-r(k)}, \ell_k) \leq 2M.$$

On the other hand $\phi(\ell_k)(\ell_k) = 1$. Therefore if $\ell_{k-r(k)} \neq \ell_k$ and there is a vertex v lying above both $\ell_{k-r(k)}$ and ℓ_k then $\Psi(v)$ can't be in the support of both

$\phi(\ell_{k-r(k)})$ and $\phi(\ell_k)$, so we get

$$\Delta_1(\ell_{k-r(k)}, \ell_k) \geq \frac{1}{C}.$$

In particular there can't be more than $N_1 = 2CM$ such indices k. A second possibility if $\ell_{k-r(k)} \neq \ell_k$ is that there is no vertex above $\ell_{k-r(k)}$ and ℓ_k, and that if v and v' denote vertices along the top above $\ell_{k-r(k)}$ and ℓ_k respectively, either $\Psi(v)$ is not in the support of $\phi(\ell_{k-r(k)})$ or $\Psi(v')$ is not in the support of $\phi(\ell_k)$. Again in this case we get

$$\Delta_1(\ell_{k-r(k)}, \ell_k) \geq \frac{1}{C}$$

so there can't be more than N_1 such indices k. The third and last possibility when $\ell_{k-r(k)} \neq \ell_k$ is that (in the same notation as the second case) the points $\Psi(v)$ and $\Psi(v')$ are in the supports of $\phi(\ell_{k-r(k)})$ and $\phi(\ell_k)$ respectively. However once the path γ is chosen, there are only finitely many intervals between the supports of distinct functions $\phi(\ell)$. Therefore there is a number N_2 independent of n such that in any ordered arrangement of points $y_1, \ldots y_r$ on γ, there can be at most N_2 indices i such that y_{i-1} is in the support of $\phi(\ell)$ and y_i is in the support of $\phi(\ell')$ with $\ell \neq \ell'$. Note that the points $\Psi(v)$ corresponding to vertices along the top, are the coordinates of the point z. Recall that this point is on the cell β. Thus the points $\Psi(v)$, as v runs through the top vertices in order, go along the path γ in order. Therefore there are at most N_2 indices j falling into the third case. These exhaust the possibilities for having $\ell_{k-r(k)} \neq \ell_k$, so that can happen at most $N = 2N_1 + N_2$ times. This proves the proposition.

Suppose η_* is a pro-chain in Z_*. Let U denote the support of η in $\mathbf{C}$, and let ξ be the supremum of $\Re x$ for $x \in U$. Fix a length L, an angular error δ, and a number σ. R will be a radius given by Proposition 6.1. Choose flows f_{ij} accordingly as described in §3.

Lemma 6.3 *Suppose L' is any number. If z is a point in the support of $F\varphi(\eta)$, $F\tau(\eta)$, $FK\varphi(\eta)$, or $F\psi(\eta)$, and $\Re g_I(z) \geq \xi - L'$, then the point z is beyond a point of η at distance $\leq (\cos\delta)^{-1}L'$.*

Proof: Suppose $z = (z_1, \ldots, z_r)$ is a point as hypothesized. Use the description in §5. There is a point $u = (u_1, \ldots, u_n) \in Z_J$ in the support of η_J, and a tree T with n top vertices, r bottom vertices, and strands labeled by numbers $t(\sigma) \in [0,1]$ and indices $i(\sigma)$ and $j(\sigma)$. We get a map $\Psi_{u,t} : T \to Z$, such that the images of the r bottom vertices are $z_1, \ldots, z_r$. This tree may be converted into a sequence of elementary arrows and a sequence of points as required in the definition of the point z being beyond the point u.

Make the convention that v_σ and v'_σ denote the top and bottom vertices of a strand σ. For any strand σ with $i(\sigma) = j(\sigma)$, then $\Psi_{u,t}(v'_\sigma) = f^0(\Psi_{u,t}(v_\sigma), s)$ for some s. For each other strand σ, let $\delta(\sigma) = D_{i(\sigma)j(\sigma)}(\Psi_{u,t}(v_\sigma), \Psi_{u,t}(v'_\sigma))$. Then z is beyond u at distance less than or equal to $M = \sum_\sigma \delta(\sigma)$. On the other hand,

$$g(z) = g(u) + \sum_\sigma (g_{i(\sigma)j(\sigma)}(\Psi_{u,t}v'_\sigma) - g_{i(\sigma)j(\sigma)}(\Psi_{u,t}v_\sigma)).$$

By hypothesis, $\Re g(u) - \Re g(z) \leq L'$. The property (3.3.2) of the flows implies that for every strand σ,

$$\delta(\sigma) \leq (\cos\delta)^{-1}(\Re g_{i(\sigma)j(\sigma)}(v_\sigma) - \Re g_{i(\sigma)j(\sigma)}(v'_\sigma)).$$

Putting these together proves the lemma.

Corollary 6.4 *Assume that there is a number r such that if $z = (z_1 \ldots, z_n) \in Supp_{Z_*}(\eta)$, then $d(z_k, P) \leq r$. There is a number R depending on r such that if z is in the support of one of the chains in the lemma, and $\Re g(z) \geq \xi - L$, then $d(z_k, P) \leq R$.*

Proof: This follows from the above lemma and a trivial modification of the proof of Proposition 6.1.

Proposition 6.5 *Suppose $\eta_* \in \sum C^{n,n}$. If z is a point in the support of $FH\varphi(\eta)_I$, then either*

$$\Re g(z) \leq \xi - L$$

or else there is a critical point ℓ in Z_I such that

$$z \in \Lambda(\ell_1) \times \ldots \Lambda(\ell_n).$$

In the latter case, there is a point z' such that $g(z)$ is in the sector $S(g(z'), \pm\delta)$, while $g(z')$ is in both of the sectors $S(g(\ell), \pm\delta)$ and $S(U, \pm\delta)$. The critical point ℓ is close enough to z' so that $|g(z') - g(\ell)| \leq (\cos\delta)^{-1}\sigma$. Furthermore the critical point is beyond points of η at distance less than $(\cos\delta)^{-1}(L + \sigma)$.

Proof: A point $z \in Z_I$ in the support of $FH\varphi(\eta)$ comes (by the process outlined in §5) from a tree T with indices $i(\sigma)$ and $j(\sigma)$ attached to the left and right of each strand; numbers $t(\sigma) \in [0,1]$ assigned to the strands; and a point $u \in Supp_{Z_*}(\eta)$. The operation H insures that the numbers assigned to the bottom strands are 1. Make a new point w by assigning numbers 0 to the bottom strands instead. Then

$$z_k = f_{i_{k-1}i_k}(w_k, 1).$$

Note that $\Re g(z) \leq \Re g(w)$.

Suppose that the first possibility does not hold, in other words $\Re g(z) \geq \xi - L$. Then also $\Re g(w) \geq \xi - L$, so by Lemma 6.3, the points z and w are beyond points of η at distance $\leq (\cos\delta)^{-1}L$. In particular, $d(z_k, P) \leq R$ and $d(w_k, P) \leq R$. Since the function g is a sum of functions each depending on one coordinate, $\Re g_{i_{k-1}i_k}(z_k) \geq \Re g_{i_{k-1}i_k}(w_k) - L$. By condition (3.3.5) satisfied by the choice of flows, this implies that there are critical points ℓ_k for $g_{i_{k-1}i_k}$ such that $z_k \in \Lambda_{i_{k-1}i_k}(\ell_k)$. So we can divide $FH\varphi$ into two pieces, the first of which is supported on $\Re g(z) \leq \xi - L$, and the second of which is supported on a union of pieces of the form $\Lambda(\ell_1, \ldots, \ell_n)$. If I has any repetitions of indices $i_{k-1} = i_k$, then $\Lambda_{i_{k-1}i_k}$ is a point, so the dimension of $\Lambda(\ell_1, \ldots, \ell_n)$ is strictly less than n. But $(FH\varphi)_I$ is a cycle of degree $n = |I|$, so any piece supported on a subset of dimension strictly less than n, vanishes. Therefore we may assume that I has no repetitions.

For each k, let t_k be the earliest time such that $f_{i_{k-1}i_k}(w_k, t_k) \in \Lambda_{i_{k-1}i_k}(\ell_k)$. Let $v_k = f_{i_{k-1}i_k}(w_k, t_k)$. If $t_k \geq 1/2$, then by condition (3.3.5) on the flows, $\Re g_{i_{k-1}i_k}(w_k) - \Re g_{i_{k-1}i_k}(v_k)$ would be at least L. This contradicts our supposition, so $t_k \leq 1/2$ for all k. But finally, condition (3.3.6) on the choice of flows says that

$$\begin{aligned} &\Re g_{i_{k-1}i_k}(f_{i_{k-1}i_k}(w_k, t_k)) - \Re g_{i_{k-1}i_k}(f_{i_{k-1}i_k}(w_k, 1)) \\ &\geq \frac{L}{\sigma} \Re g_{i_{k-1}i_k}(\ell_k) - \Re g_{i_{k-1}i_k}(f_{i_{k-1}i_k}(w_k, t_k)). \end{aligned}$$

Adding these together gives

$$\Re g_I(v_1, \ldots, v_n) - \Re g_I(z_1, \ldots, z_n) \leq \frac{L}{\sigma}(\Re g_I(\ell_1, \ldots, \ell_n) - \Re g_I(v_1, \ldots, v_n)).$$

Thus under the supposition that the first statement of the proposition does not hold, we get

$$\Re g_I(\ell_1, \ldots, \ell_n) - \Re g_I(v_1, \ldots, v_n) \leq \sigma.$$

The point $v = (v_1, \ldots, v_n)$ satisfies the requirements of the proposition. Condition (3.2.3) on the choice of the $\Lambda_{ij}(\ell)$ insures that $g(v)$ is in $S(g(\ell), \pm\delta)$, and $g(z)$ is in $S(g(v), \pm\delta)$. The same argument as in Lemma 4.5 shows that $g(v)$ is in $S(Supp_{\mathbf{C}}(\eta), \pm\delta)$. The estimate for the distance from $g(\ell)$ to $g(v)$ follows from the previous inequality and the fact about sectors. By Lemma 6.3, the point v is beyond points of η at distance less than or equal to $(\cos\delta)^{-1}L$, and the point ℓ is beyond v at distance less than $(\cos\delta)^{-1}\sigma$ (again by the equation last displayed above and the condition (3.2.4)).

Remark: Proposition 6.4 translates into a similar statement about the chain $F\tau(\eta)$. Divide $H\varphi = \alpha_1 + \alpha_2$ according to the possibilities listed in the proposition. Then set $\tau_1 = \sum_r (-KA)^r \alpha_1$ and $\tau_2 = \sum_r (-KA)^r \alpha_2$. This gives a decomposition $F\tau(\eta) = F\tau_1 + F\tau_2$, where if z is in the support of $F\tau_1$ then

$$\Re g(z) \leq \xi - L.$$

If z is in the support of $F\tau_2$ then there is a critical point ℓ in some Z_J and a point z' in $\Lambda_J(\ell)$ such that $g(z)$ is in the sector $S(g(z'), \pm\delta)$; such that $g(z')$ is in both of the sectors $S(g(\ell), \pm\delta)$ and $S(U, \pm\delta)$; such that $|g(z') - g(\ell)| \leq \sigma$; and such that the critical point ℓ is beyond points of η at a distance of less than $(\cos\delta)^{-1}L + \sigma$. These statements follow from the previous proposition and the behaviour of operations KA with respect to supports (see the proof of Lemma 4.5). It will be important in this context that $F\tau_2$ is obtained by moving $F\alpha_2$, so that our later estimates for the multiplicity of $F\alpha_2$ will yield estimates for $T\tau_2$.

7. SIZES OF CELLS

In this section we will discuss estimates for the sizes of pieces of chains like $F(KA)^r z$. We first develop some general theory of how to measure the size of a chain relative to a function, then apply it.

We should review conventions about wedge products. If V is an inner product space, then $\bigwedge^k V$ is given the norm which makes it an orthogonal quotient of $\bigotimes^k V$. Thus on a riemannian manifold, if $dx_1, \ldots, dx_k$ are an orthonormal set of cotangent vectors, then $|dx_1 \wedge \ldots \wedge dx_k| = 1$. The volume element is a form of top degree of norm one; the integral of $dx_1 \wedge \ldots \wedge dx_k$ over a unit cube is 1.

Let M be a riemannian manifold, and suppose η is a singular-de Rham chain of degree k on M. It gives a linear functional $\int_\eta \phi$ of smooth k-forms ϕ (see §4), so η can be considered as a current in the usual sense. If $\eta = h(u * H)$ for a piecewise polynomial map $h : H \to M$, then there is a constant C such that $\sup |h^*\phi| \leq C \sup |\phi|$, so η is continuous in the sup-norm. Therefore η is representable by integration in the sense of Federerer [8]. There is a positive measure $\|\eta\|$ such that

$$|\int_\eta \phi| \leq \int |\phi| \, \|\eta\|,$$

and $\|\eta\|$ is minimal for this property [8].

Suppose $r : M \to \mathbf{R}$ is a smooth function. We will consider the positive measure $r_*\|\eta\|$ on $\mathbf{R}$ defined by

$$\int_E r_*\|\eta\| = \int_{r^{-1}E} \|\eta\|$$

for $E \subset \mathbf{R}$. We introduce some notation about measures on $\mathbf{R}$. If α and β are measures, say $\alpha \prec \beta$ if for all $y \in \mathbf{R}$ we have

$$\int_{t>y} \alpha(t) \leq \int_{t>y} \beta(t).$$

Write $\alpha * \beta$ for the convolution

$$\alpha * \beta(t) = \int \alpha(s)\beta(t-s)ds.$$

For $\varepsilon > 0$ let $w(\varepsilon)$ denote the measure on $\mathbf{R}$ which is supported on $\{y \leq 0\}$ and which is given there by the L^1 function $(-y)^{\varepsilon-1}$.

Lemma 7.1 *If $\alpha_1 \prec \alpha_2$ and $\beta_1 \prec \beta_2$ then $\alpha_1 * \alpha_2 \prec \beta_1 * \beta_2$. We get the following formula for convolutions of $w(\varepsilon)$:*

$$w(\varepsilon_1) * w(\varepsilon_2) = \frac{\Gamma(\varepsilon_1)\Gamma(\varepsilon_2)}{\Gamma(\varepsilon_1 + \varepsilon_2)} w(\varepsilon_1 + \varepsilon_2).$$

Proof:

$$\begin{aligned}
\int_{t>u} \alpha_1 * \alpha_2(t) &= \int_{t>u} \int_s \alpha_1(s)\alpha_2(t-s)dsdt \\
&= \int \alpha_1(s) \int_{t>u-s} \alpha_2(t)dtds \\
&\leq \int \alpha_1(s) \int_{t>u-s} \beta_2(t)dtds \\
&= \int_{t>u} \alpha_1 * \beta_2(t)
\end{aligned}$$

so $\alpha_1 * \alpha_2 \prec \alpha_1 * \beta_2$. Similarly $\alpha_1 * \beta_2 \prec \beta_1 * \beta_2$, proving the first statement. For the second formula,

$$\begin{aligned}
w(\varepsilon_1) * w(\varepsilon_2)(y) &= \int_{u=y}^{0} (-u)^{\varepsilon_1 - 1}(u-y)^{\varepsilon_2 - 1} du \\
&= (-y)^{\varepsilon_1 + \varepsilon_2 - 1} \int_{s=0}^{1} s^{\varepsilon_1 - 1}(1-s)^{\varepsilon_2 - 1} ds \\
&= B(\varepsilon_1, \varepsilon_2) w(\varepsilon_1 + \varepsilon_2)(y),
\end{aligned}$$

refering to the beta function [36] $B(\varepsilon_1, \varepsilon_2) = \Gamma(\varepsilon_1)\Gamma(\varepsilon_2)\Gamma(\varepsilon_1 + \varepsilon_2)^{-1}$. This proves the lemma.

Remark: Inductively for a convolution of n terms we get

$$w(\varepsilon) * \ldots * w(\varepsilon) = \frac{\Gamma(\varepsilon)^n}{\Gamma(n\varepsilon)} w(n\varepsilon).$$

By Stirling's approximation [36], there is a constant $C(\varepsilon)$ such that

$$w(\varepsilon) * \ldots * w(\varepsilon) \leq n^{-\varepsilon n} C(\varepsilon)^n w(n\varepsilon).$$

Proposition 7.2 *Suppose constants C_1, C_2 and ε are given. Suppose M and N are riemannian manifolds with continuous real valued functions r and s respectively, and suppose $f : M \times [0,1] \to N$ is a continuous piecewise polynomial function. Let $|\bigwedge^k \nabla f|_0$ denote the pointwise operator norm, equal to* *the supremum of* $|\bigwedge^k \nabla f(v)|$ *over* k*-vectors* v *with* $|v| = 1$ *(this is defined separately in the various different regions where* f *is polynomial). Assume that* $|\bigwedge^k \nabla f|_0 \le C_1$*. Suppose that* $sf(x,t) \le r(x)$ *and suppose that for each* $x \in M$ *and each real number* $y > 0$ *the one dimensional measure of the set of* $f(x,t)$ *such that* $r(x) - sf(x,t) \le y$*, is less than* $C_2 y^{\varepsilon}$*. Suppose* η *is a* k*-current on* M*, and let* $\mu = r_* \|\eta\|$*. The image* $f(\eta \times [0,1])$ *is a* $k+1$*-current on* N*. Let* $\nu = s_* \|f(\eta \times [0,1])\|$*. Then*

$$\nu \prec \varepsilon C_1 C_2 w(\varepsilon) * \mu.$$

Proof: First make a reduction: assume that we have treated the case where the chain is given by a smooth form on M. Let η be a singular-de Rham k-chain. Let O_a denote a sequence of smoothing operators for currents on M, which approach the identity as $a \to 0$. If u is any continuous function on M, then [8]

$$\lim_{a \to 0} \int u \|O_a \eta\| = \int u \|\eta\|.$$

In order to define the current $f(\eta \times [0,1])$, view $\eta \times [0,1]$ as the image of a smooth current on a space H, by a piecewise polynomial map $H \to M \times [0,1]$. Then compose this with the map f to obtain an algebraic singular-de Rham chain on N.

We contend that $f(O_a \eta \times [0,1]) \rightharpoonup f(\eta \times [0,1])$ weakly as currents on N. In order to see this, we have to be more explicit about the form of O_a. There is a sequence of smooth maps $\theta_a : TM \to M$, converging to θ_0 which is just the projection p to M. The θ_a map the tangent space TM_m to a neighborhood of m, with differential equal to a times the identity at m. There is a smooth form ρ on TM supported near the zero section, such that for any $m \in M$, $\int_{TM_m} \rho = 1$. Then $O_a \eta = \theta_{a,*}(\rho * p^* \eta)$. Now suppose $\eta = h(u * H)$ for an algebraic map $h : H \to M$ and smooth form u on H (§4). We can pull back the tangent bundle of M to a bundle $H \times_M TM$ (the projections to the factors being denoted p_1 and p_2), and further multiply by the interval to get $H \times_M TM \times [0,1]$. This has pull-back forms $p_2^*(\rho)$ and $p_1^*(u)$. Then

$$O_a \eta \times [0,1] = (\theta_a p_2)(p_2^*(\rho) \wedge p_1^*(u) * H \times_M TM \times [0,1]).$$

If $f : M \times [0,1] \to N$ is a piecewise polynomial map, then

$$f(O_a \eta \times [0,1]) = (f \theta_a p_2)(p_2^*(\rho) \wedge p_1^*(u) * H \times_M TM \times [0,1]).$$

As $a \to 0$, $f\theta_1 p_2$ approaches $f\theta_0 p_2$. It does so in C^0 norm, and also, in C^1 at almost all points. Thus if φ is a smooth form on N, $(f\theta_1 p_2)^* \phi$ approaches $(f\theta_0 p_2)^* \phi$ weakly (as a distribution). So when we multiply by a smooth form on $H \times_M TM \times [0,1]$ and integrate, we get

$$\langle f(O_a \eta \times [0,1]), \phi \rangle \to \langle f(O_0 \eta \times [0,1]), \phi \rangle.$$

This proves the contention.

This weak limit of currents on N implies that

$$\liminf_{a \to 0} \int_U \|f(O_a \eta \times [0,1])\| \geq \int_U \|f(\eta \times [0,1])\|$$

for any open set $U \subset N$ [8]. Now use the assumption that we know the proposition for the smooth forms $O_a \eta$:

$$s_* \|f(O_a \eta \times [0,1])\| \prec \epsilon C_1 C_2 \omega(\epsilon) * r_* \|O_a \eta\|.$$

Let $U_y = s^{-1}\{t > y\}$. Let

$$W(x) = \int_{t > x} \omega(\epsilon)(t).$$

It is equal to zero for $x \geq 0$, and for $x < 0$ is given by

$$W(x) = \frac{(-x)^\epsilon}{\epsilon}.$$

The assumed estimate may be interpreted as saying that

$$\int_{U_y} \|f(O_a \eta \times [0,1])\| \leq \epsilon C_1 C_2 \int_M W(y - r(m)) \|O_a \eta\|(m).$$

Take the lim inf of both sides. By the two properties of the smoothing mentioned above, we get

$$\int_{U_y} \|f(\eta \times [0,1])\| \leq \epsilon C_1 C_2 \int_M W(y - r(m)) \|\eta\|(m).$$

This is the claimed inequality.

So we must treat the case when η is represented by a smooth form. In this case we can make calculations on the complement of a set of measure zero, so we may assume that the map f is smooth.

Let τ denote the coordinate on $[0,1]$. The space of $(k+1)$-covectors on $M \times [0,1]$ decomposes as an orthogonal direct sum

$$\bigwedge^{k+1} T^*(M \times [0,1])_{(m,\tau)} = (\bigwedge^{k} T^*M_m) \otimes (d\tau) \oplus \bigwedge^{k+1} T^*M_m.$$

If ϕ is a $k+1$-form on N, we can write $f^*(\phi) = \alpha_1 d\tau + \alpha_2$ according to the direct sum decomposition. Work at a point on $M \times [0,1]$, and at the corresponding image point in N. We may choose an orthonormal basis of covectors $dy_1, \ldots, dy_n$ in N such that $\partial y_i / \partial \tau = 0$ for $i \geq 2$. Thus $f^* dy_1 = (\partial y_1/\partial \tau) d\tau + \beta_1$ with $\beta_1 \in T^*M$, and $f^* dy_i \in T^*M$ for $i \geq 2$. We can write $\phi = \phi_1 dy_1 + \phi_2$ where ϕ_1 and ϕ_2 don't involve dy_1. Then

$$f^*\phi = (f^*\phi_1)(\partial y_1/\partial \tau) d\tau + (f^*\phi_1)\beta_1 + f^*\phi_2.$$

Thus $\alpha_1 = (f^*\phi_1)\partial y_1/\partial \tau$. Note that

$$|f^*\phi_1| \leq |\bigwedge^{k} \nabla f|_0 |\phi|,$$

and $|\partial y_1/\partial \tau| = |\partial f/\partial \tau|$. Therefore

$$|\alpha_1| \leq |\bigwedge^{k} \nabla f|_0 |\frac{\partial f}{\partial \tau}| f^*|\phi|.$$

By hypothesis this is less than $C_1|\frac{\partial f}{\partial \tau}| f^*|\phi|$. Now

$$\begin{aligned}
\left| \int_{f(\eta \times [0,1])} \phi \right| &= \left| \int_{\eta \times [0,1]} f^*\phi \right| \\
&= \left| \int_{\eta \times [0,1]} \alpha_1 d\tau \right| \\
&\leq \int_{M \times [0,1]} |\alpha_1| \, |\eta \times [0,1]| \\
&\leq C_1 \int_{M \times [0,1]} |\frac{\partial f}{\partial \tau}| \, \|\eta \times [0,1]\| f^*|\phi| \\
&= C_1 \int_N |\phi| f_*(|\frac{\partial f}{\partial \tau}| \, \|\eta \times [0,1]\|).
\end{aligned}$$

Therefore

$$\|f(\eta \times [0,1])\| \leq C_1 f_*(|\frac{\partial f}{\partial \tau}| \, \|\eta \times [0,1]\|).$$

Recall that U_y denotes the inverse image $s^{-1}\{t > y\}$ in N. We have

$$\begin{aligned}\int_{\{t>y\}} \nu(t) &= \int_{U_y} \|f(\eta \times [0,1])\| \\ &\leq C_1 \int_{U_y} f_*(|\frac{\partial f}{\partial \tau}| \, \|\eta \times [0,1]\|) \\ &= C_1 \int_{f^{-1}(U_y)} |\frac{\partial f}{\partial \tau}| \, \|\eta \times [0,1]\| \\ &= C_1 \int_{sf(m,\tau)>y} |\frac{\partial f}{\partial \tau}|(m,\tau) \, \|\eta\|(m).\end{aligned}$$

The last line is true because $\|\eta \times [0,1]\|$ is equal to the product of $\|\eta\|$ on M with the standard measure on $[0,1]$.

For any $m \in M$, $\int_{sf(m,\tau)>y} |\frac{\partial f}{\partial \tau}|(m,\tau)$ is equal to the one dimensional measure of the set of $f(m,\tau)$ such that $r(m) - sf(m,\tau) < r(m) - y$. Our main hypothesis says that this is equal to zero if $r(m) - y \leq 0$, and is otherwise bounded by $C_2(r(m) - y)^\epsilon$. In other words, in the notation used in the earlier reduction,

$$\int_{sf(m,\tau)>y} |\frac{\partial f}{\partial \tau}|(m,\tau) \leq \epsilon C_2 W(y - r(m)).$$

Therefore

$$\begin{aligned}\int_{t>y} \nu(t) &\leq C_1 \int_{sf(m,\tau)>y} |\frac{\partial f}{\partial \tau}|(m,\tau) \, \|\eta\|(m) \\ &\leq \epsilon C_1 C_2 \int_M W(y - r(m)) \, \|\eta\|(m) \\ &= \epsilon C_1 C_2 \int_{t>y} (\omega(\epsilon) * \mu)(t).\end{aligned}$$

This proves the proposition.

Remark: Suppose one has a map $f : M \to N$ such that $sf(m) \leq r(m)$, and $|\bigwedge^k \nabla f|_0 \leq C_1$. Then a similar (easier) argument shows that

$$s_*\|f(\eta)\| \prec C_1 r_*\|\eta\|.$$

We turn back to the situation and notation of the previous sections. We will describe two general types of estimates for chains for reference in the future, as a way of describing the estimates on the sizes of the chains constructed above. These estimates must take into account the possibility of places j in the indices

I such that $i_{j-1} = i_j$. So the estimates described below refer to the fixed choice of flow $f^0(x,t)$. For $z \in Z_I$, $t \in [0,1]$, and $1 \le j \le |I|$, define $f^0(z,t,j) \in Z_I$ by

$$f^0(z,t,j)_i = \begin{cases} z_i & \text{if } i \neq j \\ f^0(z_j,t) & \text{if } i = j \end{cases}.$$

Suppose $\eta \subset Z_I$ is an m-chain. We say that η *is degenerate for the flow* f^0 *in the coordinate* j if $i_{j-1} = i_j$ and if the chain $f^0(\eta,[0,1],j)$ is degenerate. Now we can describe the estimates.

Estimate $\mathbf{E}(\eta_n, \xi, m, C, \varepsilon, g)$:
η_n is a chain in Z_n. It is a sum of C^n pieces $\mathfrak{y}$, each in some Z_I. For each of these pieces, there are numbers $q = q(\mathfrak{y})$ and $b = b(\mathfrak{y})$ such that $b + q = m$. There are distinct $j_1, \ldots, j_b$ such that $\mathfrak{y}$ is degenerate for the flow f^0 in the coordinates $j_1, \ldots, j_b$. And finally we have estimates

$$\Re g_* |\mathfrak{y}| \prec C^n q^{-\varepsilon q} w(\varepsilon q) * \delta(\xi)$$

where $\delta(\xi)$ denotes the point mass of integral 1 at ξ.

The reason for including the notation g in the argument of the estimate $\mathbf{E}$ is that this estimate depends on rotations of g. Here is a slightly different version which does not depend on the function g.

Estimate $\mathbf{F}(\eta_n, m, C, \varepsilon)$:
η_n is a chain in Z_n. It is a sum of C^n pieces $\mathfrak{y}$, each in some Z_I. For each of these pieces, there are numbers $q = q(\mathfrak{y})$ and $b = b(\mathfrak{y})$ such that $b + q = m$. There are distinct $j_1, \ldots, j_b$ such that $\mathfrak{y}$ is degenerate for the flow f^0 in the coordinates $j_1, \ldots, j_b$. And we have estimates

$$\int |\mathfrak{y}| \prec C^n q^{-\varepsilon q}.$$

We make some simple remarks. If $\mathbf{E}(\eta_n, \xi, m, C, \varepsilon, g)$ holds and if u is a bounded form on $\mathbf{C}$ then $\mathbf{E}((g^* u) * \eta_n, \xi, m, C', \varepsilon, g)$ holds for a bigger constant $C' = C'(C, \varepsilon, w)$. Similarly for $\mathbf{F}$.

If $\Re g \ge \xi_0$ on η_n then $\mathbf{E}(\eta_n, \xi, m, C, \varepsilon, g)$ implies $\mathbf{F}(\eta_n, m, C', \varepsilon, g)$ for a bigger constant $C' = C'(C, \varepsilon, \xi, \xi_0)$.

Conversely, if $\Re g \le \xi_0$ on η_n, and $\xi > \xi_0$, then the estimate $\mathbf{F}(\eta_n, m, C, \varepsilon)$ implies $\mathbf{E}(\eta_n, \xi, m, C', \varepsilon, g)$ for a bigger C'.

The above types of estimates imply real estimates wherever I doesn't have any repetitions of indices $i_{j-1} = i_j$. More precisely, suppose η_n is a chain on Z_n. Let $Z_n^\dagger$ denote the union of the components Z_I such that $i_{j-1} \neq i_j$ for all j. Let $\eta^\dagger = \eta|_{Z^\dagger}$. Then $\mathbf{E}(\eta_n, \xi, m, C, \varepsilon, g)$ implies that

$$\Re g_* \|\eta_n^\dagger\| \prec C^n m^{-\varepsilon m} w(\varepsilon m) * \delta(\xi).$$

Similarly, $\mathbf{F}(\eta_n, m, C, \varepsilon)$ implies that

$$\int \|\eta_n^\dagger\| \prec C^n m^{-\varepsilon m}.$$

We need the more involved form described above to use inductively. When we use the estimates $\mathbf{E}$ and $\mathbf{F}$ for all n, we make the convention that ξ, C and ε do not depend on n or m. We will occasionally write $m = n/p$. In this case p also should not depend on n.

Proposition 7.3 *Given ξ and constants C and ε, there are constants C' and ε' such that for any chain $\eta = \eta_n$ in $C^{k,n}(Z_*)$ which satisfies the estimate*

$$\mathbf{E}(\eta_n, \xi, m, C, \varepsilon, g),$$

the chain $FK(AK)^r\eta$ satisfies the estimates

$$\mathbf{E}((FK(AK)^r\eta)_{n+r}, \xi, m + (r - 5n)/8, C', \varepsilon', g)$$

and

$$\mathbf{E}((FK(AK)^r\eta)_{n+r}, \xi, m, C', \varepsilon', g).$$

Proof: As usual we will allow the constants C and ε to vary (with C increasing and ε decreasing throughout the proof), but they will not depend on r, n or m. We may assume that in the definition of the estimate $\mathbf{E}$ for η, there is only one piece. Thus there are numbers b and q with $b + q = m$; there are coordinates $j_1, \ldots, j_b$ such that η is degenerate for the flow f^0 in the coordinates j_i; and we have an estimate

$$\Re g_* \|\eta_n\| \prec C^n q^{-\varepsilon q} w(\varepsilon q) * \delta(\xi).$$

By Proposition 5.1 the chain $(FK(AK)^r\eta) \cap Z_J$ is a sum of less than C^{n+r} nondegenerate chains of the form

$$\mathfrak{y} = FK_{k_r}\alpha_r \ldots K_{k_1}\alpha_1 K_{k_0}\eta.$$

We may fix one of these. Let $\eta_{-1} = \eta$, $\eta_0 = K_{k_0}\eta$, and $\eta_j = K_{k_j}\alpha_j\eta_{j-1}$, so $\mathfrak{y} = F\eta_r$. Let I^j be defined so that $\eta_j \subset Z_{I^j}$. In particular $I^0 = I$ and $I^r = J$. Consider the graph associated to $\alpha_1, \ldots, \alpha_r$ defined in §5, with an extra row of edges at the top corresponding to K_{k_0}.

Recall that the K_{k_j} determine an assignment of numbers to the edges in the graph. The number assigned to an edge is either 0, 1, or a variable s_j. We say that an edge is *marked* if it is assigned a variable s_j. There is one marked edge in each row. Say that a strand is marked if it contains a marked edge. In this case all of the other edges must be assigned the value 0. In the course of this argument, we will need to pick out many classes of edges or strands, so we will use colors as terminology. The colorings are not necessarily exclusive or exhaustive. The marking should be considered as a coloring.

There are b top vertices associated to the coordinates $j_1, \ldots, j_b$ given by the estimate E. Color the edges below these vertices *blue*. The graph represents the effect of operations performed on the chain η, and if there is a marking in one of the top strands, it means that a flow is applied to η in the corresponding coordinate. By the definition of degeneracy of η for the flow f^0 in the coordinates $j_1, \ldots, j_b$, and the fact that the flows f_{ii} are equal to f^0, the blue edges can not be marked and in fact there can be no marked edge in any strand containing a blue edge (otherwise the piece $\mathfrak{y}$ would be degenerate). Thus there are b blue strands, which are not marked.

For each non-side edge e of the graph, we have indices $l(e)$ and $r(e)$, defined as follows: if the edge e ends at the kth vertex in the jth row, then $l(e) = i^j_{k-1}$ and $r(e) = i^j_k$. (These numbers were called $i(e)$ and $j(e)$ in §§5 and 6.) The $l(e)$ and $r(e)$ are the same for all edges in a strand s, so we sometimes denote them also by $l(s)$ and $r(s)$. Say that a non-side edge e is *good* if $l(e) \neq r(e)$. Say that a strand s is *good* if $l(s) \neq r(s)$, in other words if the edges it contains are good. Note from the definition of the estimate E that any blue strand must have $l(e) = r(e)$, so it is not good.

Let p be the number of good marked strands. For each j, consider the space X_{I_j} with real valued function $\Re g \circ F$. We claim that if the marked edge in the jth row is good, then

$$(\Re g \circ F)_* \|\eta_j\| \prec C\omega(\varepsilon) * (\Re g \circ F)_* \|\eta_{j-1}\|.$$

Note first that $\|\alpha_j\eta_{j-1}\| \leq C_3(\alpha_j)_*\|\eta_{j-1}\|$ where C_3 is the bound given by (3.3.7) for the gradients of the flows in the relevant region. Thus

$$(\Re g \circ F)_* \|\alpha_j\eta_{j-1}\| \prec C_3(\Re g \circ F \circ \alpha_j)_* \|\eta_{j-1}\| = C_3(\Re g \circ F)_* \|\eta_{j-1}\|$$

since $fF\alpha_j = g\alpha_j F = gF$. Now apply Proposition 7.2 with $M = N = X_{I_j}$. This space is endowed with the real valued function $r = s = (\Re g \circ F)_*$. Since $\eta_j = K_{k_j}\alpha_j\eta_{j-1}$, we define a map $f : X_{I_j} \times [0,1] \to X_{I_j}$ by

$$f((z_1,\ldots,z_k,t_1,\ldots,t_k),\tau) = (z_1,\ldots,z_k,1,\ldots,1,\min(1,t_{k_j}+\tau),t_{k_j+1},\ldots,t_k).$$

This has the property that $\eta_j = f(\alpha_j\eta_{j-1} \times [0,1])$. The gradient of f is bounded independently of the dimension. Recall from (3.3.8) that there are constants C and ε such that for any $y > 0$ the one dimensional measure in Z of the set of τ such that $\Re g_{ij}(f_{ij}(z,t)) - \Re g_{ij}(f_{ij}(z,t+\tau)) \le y$ is less than Cy^ϵ. This shows that the hypotheses of Proposition 7.2 are satisfied. Therefore

$$\begin{aligned}(\Re g \circ F)_*\|f(\alpha_j\eta_{j-1} \times [0,1])\| &\prec \varepsilon C_4\omega(\varepsilon) * (\Re g \circ F)_*\|\alpha_j\eta_{j-1}\| \\ &\prec C_5\omega(\varepsilon) * (\Re g \circ F)_*\|\eta_{j-1}\|\end{aligned}$$

as claimed. In any case, even if the marked edge in the jth row is not good, we have

$$(\Re g \circ F)_*\|\eta_j\| \prec C(\Re g_* \circ F)_*\|\eta_{j-1}|.$$

This follows from the remark after Proposition 7.2. Finally, since $\mathfrak{y} = F\eta_r$ we have

$$\Re g_*\|\mathfrak{y}\| \prec C^{n+r}(\Re g \circ F)_*\|\eta_r\|.$$

From the initial estimate

$$\Re g_*\|\eta_{-1}\| \prec C^n q^{-\epsilon q}\omega(\varepsilon q) * \delta(\xi)$$

and the series of estimates given above, we get

$$\Re g_*\|\mathfrak{y}\| \prec C^{n+r}p^{-\epsilon p}q^{-\epsilon q}w(\varepsilon p) * w(\varepsilon q) * \delta(\xi).$$

Now $p^{-\epsilon p}q^{-\epsilon q}w(\varepsilon p) * w(\varepsilon q) \prec C^{p+q}(p+q)^{-\epsilon(p+q)}w(\varepsilon(p+q))$ by Lemma 7.1 and Stirling's approximation, so

$$\Re g_*\|\mathfrak{y}\| \prec C^{n+r}(p+q)^{-\epsilon(p+q)}w(\varepsilon(p+q)) * \delta(\xi).$$

Thus $p+q$ will play the role of q in the new estimate **E** we are looking for.

Let $j'_1,\ldots,j'_a$ be the coordinates in which $\mathfrak{y}$ is degenerate for the flow f^0. Color the corresponding strands *yellow*. The number a will play the role of b in the new estimate **E**. Thus we have to prove that $a+p+q \ge m+(r-5n)/8$, or in other words that

$$p+a \ge b+(r-5n)/8.$$

For the second estimate, we have to prove that $p + a \geq b$.

Any marked strand s in the bottom row such that $l(s) = r(s)$ must be one of the yellow strands, because applying the flow f^0 in that coordinate results in two applications of f^0 to the same point, and hence in a degenerate cell. Sub-color the yellow strands which arise in this fashion *orange-yellow*. We will identify another reason for a strand to be yellow below, and color those strands *green-yellow*. We will show that the total number of strands which are either good marked strands, orange-yellow strands, or green-yellow strands, is at least $b + (r - 5n)/8$.

The tree T composed of the strands (including side strands) is a disjoint union of $n + 2$ trees T_i, each of which is a binary tree with a single top strand coming from one of the top vertices. There are r forks in T, $n + 2$ top strands, $n + r + 2$ bottom strands, and $2r + n + 2$ strands altogether. In particular there are r strands which are not bottom strands. There are $r + 1 \geq r$ marked strands (Propositions 7.5 and 7.7 below will have the same proof as the present one, but with r marked edges instead of $r + 1$; with this in mind, we only use the fact that the number of marked edges is $\geq r$), but these do not include any of the b top strands colored blue. Therefore there are at least b marked strands at the bottom.

Notice first of all that any marked strand at the bottom is either a good marked strand, or is orange-yellow. Therefore all of the marked strands on the bottom contribute to our count. In particular, we get $p + a \geq b$, which proves the second estimate. Choose b of the bottom marked strands, remove the marking or coloring, and color them *black*. Now there are $\geq r - b$ marked strands left. We must show that the number of good marked strands, orange-yellow strands, green-yellow strands, and black strands is at least $b + (r - 5n)/8$, or equivalently that the number of good marked strands, orange-yellow strands, and green-yellow strands left is at least $(r - 5n)/8$.

If there are $(r - 5n)/8$ marked strands at the bottom, then we are done, so we may suppose that the number of marked strands at the bottom is less than or equal to $(r - 5n)/8 - 1$.

Color some of the forks *red* as follows. A fork is to be colored red if it is attached to at least one good marked strand, orange-yellow strand, or green-yellow strand. The number of strands we are counting is at least half the number of red forks (a strand could end in two red forks), so we have to show that the number of red forks is at least $(r - 5n)/4$.

Say that a fork is *good* if for the three strands s_1, s_2, s_3 coming out from it, it is not the case that $r(s_1) = r(s_2) = r(s_3) = l(s_1) = l(s_2) = l(s_3)$. If a fork is good then at least two of the strands coming from it will be good. A good fork will be colored red if there are at least two marked strands coming from it, for in this case at least one of those will be a good marked strand.

If the cell $\mathfrak{y}$ is not degenerate, then any fork with three marked strands must be good and hence colored red. This is because if a fork is not good, then the flows associated to the strands are all the same, say $f(x,t)$. If x is a point in Z, the 3-cell composed of points

$$(f(f(x,t_1),t_2),f(f(x,t_1),t_3))$$

in $Z \times Z$ is degenerate, being contained in the product of two copies of the one-dimensional set of points $f(x,t)$.

Suppose a fork has two marked strands attached, and suppose it has a bottom strand attached. If the bottom strand is one of the marked ones, then it is either a good marked strand or an orange-yellow strand. In either case, the fork is red. On the other hand, suppose that the two marked strands are not bottom strands, but the third strand is a bottom one. If the fork is good, then it is colored red. If it is not good, then we claim that the bottom strand is a yellow strand. The yellow strands which arise this way are to be colored *green-yellow* (these are the ones refered to above). To prove the claim, let j denote the location the bottom strand in question. Consider the chain $f^0(\mathfrak{y},[0,1],j)$. It is obtained from Y by the graph obtained from the original graph by assigning to the strand in question the variable $s \in [0,1]$. But now in this graph, there are three marked strands attached to one fork, all of which have $l(s) = r(s)$. Thus the resulting chain is degenerate as explained in the previous paragraph. Therefore $\mathfrak{y}$ is degenerate for the flow f^0 in the jth coordinate, so the jth bottom strand is yellow as claimed. So, whether good or not, the fork is colored red.

To summarize, any fork which has three marked strands is colored red, and any fork which has two marked strands and is attached to a bottom strand is colored red. We now give a lower bound for the number of red forks which arise for these two reasons.

Let t_i be the number of strands at the bottom of T_i, r_i the number of marked strands in T_i, and q_i the number of marked strands which are not at the bottom. We have

$$\sum_i t_i = n + r + 2$$

$$\sum_i r_i \geq r - b$$

$$\sum_i q_i \geq r - b - (r - 5n)/8.$$

The second equation is because in recoloring b strands black, we have removed the marking from no more than b of the originally marked strands. The last equation then follows from the hypothesis that no more than $(r - 5n)/8 - 1$ bottom strands are marked. Note that $b \leq n$, so we get

$$\sum_i q_i \geq 7r/8 - 3n/8 + 1.$$

Let T_i' be the sub-tree of T_i which is the complement of the set of bottom strands which emanate from forks with two bottom strands. Let s_i be the number of such forks, so $2s_i$ strands have been eliminated. In particular, $s_i \leq t_i/2$. The number of bottom strands of T_i' is $t_i - s_i$. Mark also the strands in T_i' which are bottom strands in T_i. Now the number of marked strands in T_i' is at least $q_i + t_i - 2s_i$. The forks in T_i' which have three marked strands are colored red in T_i for the two above described reasons. Let y_i be the number of forks with three marked strands in T_i'. We have to show that $\sum_i y_i \geq (r - 5n)/4$. The number of forks in T_i' is $t_i - s_i - 1$. The number of end strands, including the one at the top, is $t_i - s_i + 1$. The number of forks with ≤ 2 marked strands is $t_i - s_i - 1 - y_i$. The number of marked strands at the ends is less than or equal to $t_i - s_i + 1$. Therefore the number of marked strands is less than or equal to

$$\frac{1}{2}(3y_i + 2(t_i - s_i - 1 - y_i) + t_i - s_i + 1).$$

In other words,

$$\begin{aligned} 2(q_i + t_i - 2s_i) &\leq 3y_i + 2(t_i - s_i - 1 - y_i) + t_i - s_i + 1 \\ &= y_i + 3t_i - 3s_i - 1 \end{aligned}$$

or

$$y_i \geq 2q_i - t_i - s_i + 1.$$

Now recall that $s_i \leq t_i/2$, so

$$y_i \geq 2q_i - 3t_i/2 + 1.$$

Adding this up over i we get

$$\begin{aligned} \sum_i y_i &\geq 2(7r/8 - 3n/8 + 1) - 3(n + r + 2)/2 + n + 2 \\ &= 14r/8 - 6n/8 - 12r/8 - 12n/8 + 8n/8 + 1 \\ &\geq 2r/8 - 10n/8 = (r - 5n)/4 \end{aligned}$$

as required. This completes the proof of the proposition.

Corollary 7.4 *Given ξ and constants p, C and ε, there are constants p', C' and ε' such that for any chain $\eta = \eta_n$ which satisfies the estimate*

$$\mathbf{E}(\eta_n, \xi, n/p, C, \varepsilon, g),$$

the chain $FK(AK)^r\eta$ satisfies the estimate

$$\mathbf{E}((FK(AK)^r\eta)_{n+r}, \xi, (n+r)/p', C', \varepsilon', g).$$

Proof: Using Proposition 7.3, we must choose p' so that

$$(n+r)/p' \leq \max(n/p, n/p + (r-5n)/8).$$

Set $p' = \max(7p, 56)$. Then if $n \leq r/6$ we get $r - 5n \geq r/6 \geq (n+r)/7$, so $(r-5n)/8 \geq (n+r)/56 \geq (n+r)/p'$. On the other hand if $n \geq r/6$ then $n + r \leq 7n$ so $n/p \geq (n+r)/7p \geq (n+r)/p'$.

Proposition 7.5 *Given ξ and constants C and ε, there are constants C' and ε' such that for any chain $\eta = \eta_n$ which satisfies the estimate*

$$\mathbf{E}(\eta_n, \xi, m, C, \varepsilon, g),$$

the chain $FH(AK)^r\eta$ satisfies the estimates

$$\mathbf{E}((FH(AK)^r\eta)_{n+r}, \xi, m + (r-5n)/8, C', \varepsilon', g)$$

and

$$\mathbf{E}((FH(AK)^r\eta)_{n+r}, \xi, m, C', \varepsilon', g).$$

Proof: The proof is exactly the same as the proof of Proposition 7.3, except that in the graph associated to a cell

$$FH\alpha_r K_{k_{r-1}} \ldots \alpha_1 K_{k_0}\eta,$$

there are only r marked strands instead of $r+1$. In the proof of 7.3 we only used the fact that the number of marked strands was $\geq r$.

Corollary 7.6 *Given ξ and constants p, C and ε, there are constants p', C' and ε' such that for any chain $\eta = \eta_n$ which satisfies the estimate*

$$\mathbf{E}(\eta_n, \xi, n/p, C, \varepsilon, g),$$

the chain $FH(AK)^r\eta$ satisfies the estimate

$$\mathbf{E}((FH(AK)^r\eta)_{n+r}, \xi, (n+r)/p', C', \varepsilon', g).$$

Proof: The same as for Corollary 7.4.

Proposition 7.7 *Under the above assumption, given ξ and constants C and ε, there are constants C' and ε' such that for any chain $\phi_n \subset X_n$ such that $H\phi_n$ satisfies the estimate*

$$\mathbf{E}(H\phi_n, \xi, m, C, \varepsilon, g),$$

(explained in the previous remark), the chain $F(KA)^r H\phi$ satisfies the estimates

$$\mathbf{E}((F(KA)^r H\phi)_{n+r}, \xi, m + (r - 5n)/8, C', \varepsilon', g)$$

and

$$\mathbf{E}((F(KA)^r H\phi)_{n+r}, \xi, m, C', \varepsilon', g).$$

Proof: This is essentially the same as the proof of Proposition 7.3. $F(KA)^r H\phi$ is composed of cells

$$\mathfrak{y} = FK_{k_r}\alpha_r \ldots K_{k_1}\alpha_1 H\phi,$$

obtained from ϕ by graphs with all top edges assigned the value 1, and with r marked edges. Equivalently, we can think of the $\mathfrak{y}$ as obtained from $FH\phi$ by the same graphs, but with the top strands assigned the value 0. In following the proof of 7.3, this second point of view may be more useful. We may assume $\phi \subset X_I$. The chain $F(KA)^r H\phi$ will be degenerate if there is any j such that $i_{j-1} = i_j$. Therefore the number b in the estimate for $H\phi$ is zero, in other words we have an estimate

$$\Re g_* \|FH\phi_n\| \prec C^n m^{-\varepsilon m} w(\varepsilon m) * \delta(\xi).$$

In the notation of the proof of 7.3, $\eta_{-1} = FH\phi$, so

$$\Re g_* \|\eta_{-1}\| \prec C^n m^{-\varepsilon m} w(\varepsilon m) * \delta(\xi).$$

Thus if p is the number of good marked edges, then

$$\Re g_* \|\mathfrak{y}\| \prec C^n (p+m)^{-\varepsilon(p+m)} w(\varepsilon(p+m)) * \delta(\xi).$$

So again if a is the number of coordinates for which $\mathfrak{y}$ is degenerate for the flow f^0, we must prove that $p + m + a \geq m + (r - 5n)/8$, in other words that $p + a \geq (r - 5n)/8$ (in this case $p + m + a \geq m$ is obvious). From here the proof is exactly the same as the proof of 7.3, using the fact that there are at least r marked edges as in 7.5.

Corollary 7.8 *Given ξ and constants p, C and ε, there are constants p', C' and ε' such that for any chain $\phi_n \subset X_n$ which satisfies the estimate*

$$\mathbf{E}(H\phi_n, \xi, n/p, C, \varepsilon, g),$$

the chain $F(KA)^r H\phi$ satisfies the estimate

$$\mathbf{E}((F(KA)^r H\phi)_{n+r}, \xi, (n+r)/p', C', \varepsilon', g).$$

Proof: The same as for Corollary 7.4.

8. Moving the cycle of integration

We will now apply the procedure outlined in the previous sections to move the cycle of integration η to obtain an analytic continuation.

Inductive hypothesis: Suppose that ζ_0 is a point in $\mathbf{C}$, with a path ρ_0 from $|\zeta| \geq a$ to ζ_0, of length $\leq M_0$, not meeting S_{M_0}. Suppose that η_* is a prochain with $(\partial - A)\eta_* = 0$, such that all points of $Supp_{Z_*}(\eta_*)$ are beyond points of $Supp_{Z_*}(\beta_*)$ at distance $\leq M_0$, and such that there is an estimate $\mathbf{F}(\eta_n, \varepsilon n, C, \varepsilon)$ with C and ε uniform for all η_n. Suppose that ζ_0 is not contained in $Supp_{\mathbf{C}}(\eta_*)$, and that

$$f(\zeta) = \int_\eta \frac{b}{g - \zeta}$$

serves to define the analytic continuation of f along the path ρ_0, for ζ near ζ_0. Finally, assume that the path ρ_0 is piecewise linear, and that the chain η is obtained by repeated applications of the procedure we are about to outline.

Fix a number L and let $M = M_0 + L$. Suppose $\rho : [0,1] \to \mathbf{C}$ is a line segment of length L, which does not meet S_M, and which begins at $\rho(0) = \zeta_0$. We would like to analytically continue $f(\zeta)$ along the segment ρ. Without loss of generality, we may make a rotation and assume that the segment points in the negative real direction, in other words $\rho(1) = \zeta_0 - L$.

Choose a small number ϵ such that the disc $D(\zeta_0, 10\epsilon)$ does not meet $Supp_{\mathbf{C}}(\eta_*)$, and such that the neighborhood $D(\rho, 10\epsilon)$ (signifying the set of all points at distance less than 10ϵ from the segment ρ) does not meet S_M. Choose numbers σ and δ, small enough to meet the requirements made below. Let $L' = L - \epsilon$. Let $\xi_0 = \Re\zeta_0$.

Make a choice of flows as in §3, with reference to this angular error δ, the length L' (which is slightly shorter than L), and the small number σ. Remember that a rotation has been made, so the flows should be chosen to go approximately in the negative real direction. In terms of a picture fixed from the start, the flows would go in some direction approximately equal to the direction of the line segment ρ. The flow f^0 must be fixed independently of which rotation is made, because it appears in the definition of the estimate $\mathbf{F}$, hence in the inductive hypothesis.

Divide the chain η into two pieces $\eta = \eta_1 + \eta_2$, such that η_1 is supported in $D(\rho, 3\epsilon)$ and η_2 is supported outside of $D(\rho, 2\epsilon)$. To do this, let $u(x)$ be a cutoff function so that $u(x) = 1$ for $x \in D(\rho, 2\epsilon)$ and $u(x) = 0$ for x outside of $D(\rho, 3\epsilon)$. Set

$$\eta_1 = (g^* u)\eta, \qquad \eta_2 = (1 - g^* u)\eta.$$

The boundary of η_1 is

$$\nu = (\partial - A)\eta_1 = (g^* du) * \eta,$$

which is supported in the annular region

$$A(\rho, 2\epsilon, 3\epsilon) = D(\rho, 3\epsilon) - D(\rho, 2\epsilon).$$

Since the support of η does not meet $D(\zeta_0, 10\epsilon)$, neither do the supports of η_1, η_2, or ν. Thus, for example, the support of ν is contained in the U-shaped region $A(\rho, 2\epsilon, 3\epsilon) - D(\zeta_0, 10\epsilon)$ and the support of η_1 is contained in $D(\rho, 3\epsilon) - D(\zeta_0, 10\epsilon)$. If $x \in Supp_{\mathbf{C}}(\eta_1)$ then $\Re x \leq \xi_0 - 5\epsilon$.

Construct chains $FK\varphi(\eta_1)$, $F\tau(\eta_1)$, and $F\psi(\eta_1)$ as described in §4. They have the following properties.

Equation: Apply F to the equation of Lemma 4.4. Since A commutes with F, and η_* is a chain on Z_*,

$$(\partial + A)FK\varphi = F\tau - F\psi - \eta_1.$$

Estimates for sizes: The chains η_1, η_2, and ν satisfy estimates of the form $\mathbf{F}(\cdot, \varepsilon n, C, \varepsilon)$. Since $Supp_{\mathbf{C}}(\eta_1)$ and $Supp_{\mathbf{C}}(\nu)$ are contained in the region $\Re x \leq \xi_0 - 5\epsilon$, these estimates translate into

$$\mathbf{E}(\eta_{1,n}, \xi_0, \varepsilon n, C, \varepsilon, g)$$

$$\mathbf{E}(\nu_n, \xi_0, \varepsilon n, C, \varepsilon, g).$$

Here the constant C has been increased, but is still independent of n. By the Propositions and Corollaries 7.3–7.7, we can decrease ε and increase C so as to obtain estimates

$$\mathbf{E}((FK\varphi)_n, \xi_0, \varepsilon n, C, \varepsilon, g)$$

$$\mathbf{E}((F\tau)_n, \xi_0, \varepsilon n, C, \varepsilon, g)$$

$$\mathbf{E}((F\psi)_n, \xi_0, \varepsilon n, C, \varepsilon, g).$$

These are relative estimates. After the chains are cut off further (see below), so that they are supported on $\Re x \geq \xi_0 - L - 5\epsilon$, then the relative estimates $\mathbf{E}$ will imply absolute estimates of the form $\mathbf{F}$.

Estimates for supports: Recall that $S(x, \pm\delta)$ denotes the sector opening in the negative real direction from the point x, with angular opening 2δ. If U is a subset of $\mathbf{C}$, then $S(U, \pm\delta)$ denotes the union of sectors $S(x, \pm\delta)$ for $x \in U$. The supports of $FK\varphi$ and $F\tau$ are contained in

$$S(D(\rho, 3\epsilon) - D(\zeta_0, 10\epsilon), \pm\delta)$$

while the support of $F\psi$ is contained in

$$S(A(\rho, 2\epsilon, 3\epsilon) - D(\zeta_0, 10\epsilon), \pm\delta).$$

If the angle δ is chosen small enough, then this last region does not meet $D(\rho, \epsilon)$. Neither region meets $D(\zeta_0, 9\epsilon)$.

The support of $F\tau$: As in the remark following Corollary 6.4, the chain $H\varphi$ is a sum of pieces $H\varphi = \alpha_1 + \alpha_2$ with the following properties of supports. If $z \in Supp_{Z_*}(F\alpha_1)$ then

$$\Re g(z) \leq \xi_0 - 5\epsilon - L' = \xi_0 - 4\epsilon - L.$$

On the other hand, $Supp_{Z_*}(F\alpha_2)$ is contained in a union of subsets of the form $\Lambda(\ell)$, for critical points ℓ of g on Z_I. If $z \in Supp_{Z_*}(F\alpha_2) \cap \Lambda(\ell)$ with $\Re g(z) \geq \xi_0 - 5\epsilon - L'$ then there is a point z' in $\Lambda(\ell)$ such that

$$|g(\ell) - g(z')| \leq (\cos\delta)^{-1}\sigma,$$

and

$$g(z') \in S(g(\ell), \pm\delta) \cap S(Supp_{\mathbf{C}}(\eta_1), \pm\delta)$$
$$g(z) \in S(g(z'), \pm\delta).$$

Furthermore, in this situation there is a point w in $Supp_{Z_*}(\eta_1)$ such that ℓ is beyond w at distance less than $(\cos\delta)^{-1}(L' + \sigma)$. If δ and σ are small, then this is less than L (since $L' = L - \epsilon$). Thus ℓ is beyond points of β_* at distance less than M, so $g(\ell) \in S_M$ by the definition of S_M.

Now use the assumption that S_M does not contain any point in $D(\rho, 10\epsilon)$. It implies that $g(\ell)$ is not contained in $D(\rho, 10\epsilon)$. On the other hand, we may assume $(\cos\delta)^{-1}\sigma \leq \epsilon$, so $g(z')$ is not contained in $D(\rho, 9\epsilon)$. But

$$S(Supp_{\mathbf{C}}(\eta_1), \pm\delta) - D(\rho, 9\epsilon)$$

is contained in the region $\Re x \leq \xi_0 - 4\epsilon - L$, so we finally conclude that if $z \in Supp_{Z_*}(FH\varphi)$, then $\Re g(z) \leq \xi_0 - 4\epsilon - L$. Now $F\tau = \sum_r F(-KA)^r H\varphi$, so

$$Supp_{\mathbf{C}}(F\tau) \subset S(Supp_{\mathbf{C}} FH\varphi, \pm\delta).$$

Therefore $Supp_{\mathbf{C}}(F\tau)$ is contained in the region $\Re x \leq \xi_0 - 4\epsilon - L$. This is the statement which essentially says that we have moved the chain of integration away from the segment ρ.

The new cycle of integration: Let v be another cutoff function, with $v(x) = 0$ for $\Re x \leq \xi_0 - 4\epsilon - L$, and $v(x) = 1$ for $\Re(x) \geq \xi_0 - 3\epsilon - L$. Set

$$\begin{aligned} \eta' &= \eta + (\partial + A)(g^* v)FK\varphi \\ &= \eta_1 + \eta_2 + (g^* dv) * FK\varphi + (g^* v)F\tau - (g^* v)F\psi - (g^* v)\eta_1. \end{aligned}$$

Now $F\tau$ is supported in $\Re x \leq \xi_0 - 4\epsilon - L$, so $(g^* v)F\tau = 0$. On the other hand, η_1 is supported in $D(\rho, 3\epsilon)$, which is contained in the region $\Re x \geq \xi_0 - 3\epsilon - L$ (note that $\Re\rho(1) = \xi_0 - L$). Therefore $(g^* v)\eta_1 = \eta_1$, which cancels the other η_1 term. Thus

$$\eta' = \eta_2 + (g^* dv) * FK\varphi - (g^* v)F\psi.$$

All of these terms are supported in the complement of $D(\rho, \epsilon)$, the first by the definition of η_2, the second because dv is supported in $\xi_0 - 4\epsilon - L \leq \Re x \leq \xi_0 - 3\epsilon - L$, and the third because $Supp_{\mathbf{C}}(F\psi)$ is contained in $S(Supp_{\mathbf{C}}(\nu), \pm\delta)$. In the third case the support of ν is contained in a U-shaped region which misses $D(\rho, 2\epsilon)$ so the sector will miss $D(\rho, \epsilon)$ if δ is small. So $Supp_{\mathbf{C}}(\eta')$ does not meet the neighborhood $D(\rho, \epsilon)$ of the segment ρ.

Let $\eta'' = \eta' - \eta_2$. Multiplying $FK\varphi$ and $F\psi$ by bounded forms preserves the estimates $\mathbf{E}$, so we have the estimate $\mathbf{E}(\eta''_n, \xi_0, \varepsilon n, C, \varepsilon, g)$. As usual, the constant C may be increased, but it remains uniform in n. But $Supp_{\mathbf{C}}(\eta'')$ is contained in $\Re x \geq \xi_0 - 4\epsilon - L$, so the relative estimate $\mathbf{E}$ implies the absolute estimate $\mathbf{F}(\eta''_n, \varepsilon n, C, \varepsilon)$. We know the same estimate for η_2, so we obtain the estimate $\mathbf{F}(\eta'_n, \varepsilon n, C, \varepsilon)$. Now if I is an index with any repetition $i_{k-1} = i_k$, then the differential form b_I is equal to zero (this is where we use the fact that the matrix B has diagonal entries equal to zero). On the other hand, if the index I (with $|I| = n$) contains no repetitions, then the estimate $\mathbf{F}(\eta'_I, \varepsilon n, C, \varepsilon)$ translates into an actual estimate for sizes:

$$\int_{Z_I} \|\eta'_I\| \leq C^n n^{-\varepsilon n}.$$

If ζ is in the neighborhood $D(\rho, \epsilon/2)$, then $|(g - \zeta)^{-1}| \leq 2/\epsilon$ on the support of η', so the sum of integrals

$$\int_{\eta'} \frac{b}{g - \zeta} = \sum_I \int_{\eta'_I} \frac{b_I}{g_I - \zeta}$$

converges with estimates as required in condition (2.5.2). There is a homology $\kappa = (g^*v)FK\varphi$ such that $\eta' = \eta + (\partial + A)\kappa$, and such that $Supp_{\mathbf{C}}(\kappa)$ does not meet $D(\zeta_0, \epsilon)$. Thus we may apply Lemma 4.3 to conclude that for $\zeta \in D(\zeta_0, \epsilon/2)$,

$$\int_{\eta'} \frac{b}{g - \zeta} = \int_{\eta} \frac{b}{g - \zeta} = f(\zeta).$$

Therefore

$$f(\zeta) = \int_{\eta'} \frac{b}{g - \zeta}$$

defines the analytic continuation of $f(\zeta)$ in a neighborhood of the line segment ρ.

Inductive hypothesis: We have to check that the new chain η' satisfies our inductive hypothesis, so the procedure can be applied over again. Note that $(\partial - A)\eta' = (\partial - A)(\partial + A)\kappa = 0$ since ∂ and A commute, and $A^2 = 0$ by Lemma 4.2. The estimate $\mathbf{F}(\eta'_n, \varepsilon n, C, \varepsilon)$ holds. We just have to show that points of η'_* are beyond points of β_* at distance $\leq M = M_0 + L$. By transitivity and the inductive hypothesis for η, it suffices to show that points of η' are beyond points of η at distance $\leq L$. Clearly this is already true for points in the support of η_2. Suppose z is a point on the support of η''. Then $\Re g(z) \geq \xi_0 - 5\epsilon - L'$, and z is either in the support of $FK\varphi$ or $F\psi$. The supremum of $\Re g(w)$, for w in the support of η_1 or ν, is less than $\xi_0 - 5\epsilon$. We may apply Lemma 6.3 to conclude that the point z is beyond a point in the support of η at distance less than or equal to $(\cos\delta)^{-1}L' \leq L$.

This completes the description of the inductive procedure for analytically continuing $f(\zeta)$ in the complement of the set of singularities. It shows that condition (2.5.2) holds, so $f(\zeta)$ has an analytic continuation with locally finite branching.

Now let us consider how to modify the argument to obtain information near a singularity. The remainder of this section will contain a description of the process for proving condition (2.5.3). The proof will depend on an estimate of the multiplicities of chains $FH\varphi$ on sets $\Lambda(\ell)$, to be proved in the next section.

Suppose that $f(\zeta)$ has been analytically continued to a neighborhood of a point ζ_0 by successive applications of the above process. Preserve the notations and assumptions of the above inductive hypothesis. Suppose $M > M_0$ and s is a point in S_M, and suppose that there is a line segment $\rho : [0,1] \to \mathbf{C}$ such that $\rho(0) = \zeta_0$ and $\rho(1/2) = s$. As usual, we may rotate and assume that ρ goes in the negative real direction. Let L denote the length of the segment ρ, so

$L = 2|\zeta_0 - s|$. Suppose that $L + M_0 \leq M$, and suppose that s is the only point of S_M contained in the segment ρ. Choose a small number ϵ so that $D(\zeta_0, 10\epsilon)$ does not intersect $Supp_{\mathbf{C}}(\eta)$, and so that s is the only point of S_M contained in $D(\rho, 10\epsilon)$. Apply the same procedure as described above, obtaining chains $FK\varphi$, $F\tau$, and $F\psi$. The previous discussion applies equally well, until the point in the treatment of the support of $FH\varphi$ where it is assumed that S_M does not meet $D(\rho, 10\epsilon)$. Let us pick up the discussion from there.

Recall that $H\varphi = \alpha_1 + \alpha_2$, where $F\alpha_1$ was supported on $\Re g(z) \leq \xi_0 - 4\epsilon - L$. On the other hand, $F\alpha_2$ was supported on a union of sets of the form $\Lambda(\boldsymbol{\ell})$ for critical points $\boldsymbol{\ell} \in Z_I$ with $g(\boldsymbol{\ell}) \in D(\rho, 10\epsilon)$. The critical points $\boldsymbol{\ell}$ which occur are all beyond points of η at distance less than L, so they are beyond points of β_* at distance less than M. By definition, $g(\boldsymbol{\ell}) \in S_M$. In this case, there is only one point s in both S_M and $D(\rho, 10\epsilon)$, so $g(\boldsymbol{\ell}) = s$. By Condition (3.2.3) on the choice of sets Λ, the chain $F\alpha_2$ is supported in $g(z) \in S(s, \pm\delta)$. Let $\xi_1 = \Re s = \xi_0 - L/2$, so in particular $F\alpha_2$ is supported in $\Re g(z) \leq \xi_1$.

The dimension of the chain $(F\alpha_2)_n$ (the restriction of $F\alpha_2$ to Z_n) is equal to n, and this is also the real dimension of the set $\Lambda(\boldsymbol{\ell})$. So the chain $F\alpha_2$ is just the fundamental class of $\Lambda(\boldsymbol{\ell})$, multiplied by some function. This function gives the *multiplicity* of $F\alpha_2$ at a given point of $\Lambda(\boldsymbol{\ell})$. If $\boldsymbol{\ell} = (\ell_1, \ldots, \ell_n) \in Z_I$ is a critical point occuring, then the index I has no repetitions (otherwise the dimension of $\Lambda(\boldsymbol{\ell})$ would be too small), and $d(\ell_k, P) \leq R$. The number of such critical points in Z_n is bounded by C^n.

Proposition 8.1 *The estimate* $\mathbf{E}((F\alpha_2)_n, \xi_1, \varepsilon n, C, \varepsilon, g)$ *holds.*

Proof: The main step will be treated in the next section. It says that if $\boldsymbol{\ell}$ is a critical point in Z_I with $|I| = n$, then the multiplicity of $F\alpha_2$ at any point in $\Lambda(\boldsymbol{\ell})$ is bounded by C^n. Let us assume for the moment that this is true.

Suppose that if $\boldsymbol{\ell}$ is a critical point in Z_I with $|I| = n$, and $g(\boldsymbol{\ell}) = s$. If $\lambda = (\lambda_1, \ldots, \lambda_n)$ is a point in $\Lambda(\boldsymbol{\ell}) = \Lambda(\ell_1) \times \cdots \times \Lambda(\ell_n)$, then

$$\Re g(\lambda) = \xi - \sum_{k=1}^{n} (\Re g_{i_{k-1} i_k}(\ell_k) - \Re g_{i_{k-1} i_k}(\lambda_k)).$$

For any k, $i_{k-1} \neq i_k$ since there are no repetitions of indices in I, so condition (3.2.5) on the choices of the sets $\Lambda_{i_{k-1} i_k}(\ell_k)$ implies that

$$\Re g_{i_{k-1} i_k}(\ell_k) - \Re g_{i_{k-1} i_k}(\lambda_k) \geq c(d_\Lambda(\ell_k, \lambda_k))^\nu.$$

Here d_Λ denotes the linear distance in $\Lambda(\ell_k)$. We can reinterpret this condition in the notation of §7 as saying that

$$(\Re g_{i_{k-1}i_k})_* \|\Lambda_{i_{k-1}i_k}(\ell_k)\| \prec C\omega(\varepsilon) * \delta(g_{i_{k-1}i_k}(\ell_k))$$

for small ε. The direct image of a product of positive measures, by a function which is the sum of the functions on the components, is equal to the convolution of the individual direct images of the component measures. Therefore

$$\Re g_* \|\Lambda(\ell)\| \prec C^n n^{-\varepsilon n} \omega(\varepsilon n) * \delta(\xi_1).$$

Now the bound on the multiplicity, together with the fact that at most C^n critical points ℓ can occur in Z_n, imply that

$$\Re g_* \|F\alpha_{2,n}\| \prec C^n n^{-\varepsilon n} \omega(\varepsilon n) * \delta(\xi_1).$$

Since the indices I which can occur have no repetitions, this is an estimate of the form $\mathrm{E}((F\alpha_2)_n, \xi_1, \varepsilon n, C, \varepsilon, g)$.

Recall that $\tau = \sum_r (-KA)^r H\varphi$. Write $\tau = \tau_1 + \tau_2$ with $\tau_i = \sum_r (-KA)^r \alpha_i$. Then $F\tau_1$ is supported in $\Re x \leq \xi_0 - 4\epsilon - L$, and $F\tau_2$ is supported in $S(s, \pm\delta)$.

Corollary 8.2 *The chain $F\tau_2$ satisfies an estimate of the form*

$$\mathrm{E}((F\tau_2)_n, \xi_1, \varepsilon n, C, \varepsilon, g).$$

Proof: This follows immediately from the above proposition and the statements given at the end of §7.

We can now finish the proof of condition (2.5.3). Use a cutoff function v as it was used previously, setting

$$\begin{aligned} \eta' &= \eta + (\partial + A)(g^*v)FK\varphi \\ &= \eta_1 + \eta_2 + (g^*dv) * FK\varphi + (g^*v)F\tau_1 + (g^*v)F\tau_2 - (g^*v)F\psi - (g^*v)\eta_1. \end{aligned}$$

As before, $(g^*v)F\tau_1 = 0$ and $(g^*v)\eta_1 = \eta_1$, so

$$\eta' = \eta_2 + (g^*dv) * FK\varphi + (g^*v)F\tau_2 - (g^*v)F\psi.$$

Set $\eta'' = \eta_2 + (g^*dv) * FK\varphi - (g^*v)F\psi$, so $\eta' = \eta'' + (g^*v)F\tau_2$.

The integral

$$f(\zeta) = \int_{\eta'} \frac{b}{g - \zeta}$$

converges when ζ is in $D(\rho, \epsilon)$ but not in $S(s, \pm\delta)$. It represents the analytic continuation of $f(\zeta)$ toward the singularity along the segment ρ. Let T be a pie-slice opening from the singularity s in the positive real direction, with angular opening 2δ and radius $L/4$.

Pick a number k and restrict to values of n such that $\varepsilon n - 2 \geq k$. Then for $\zeta \in T$,

$$\frac{1}{k!}\frac{d^k}{d\zeta^k} f(\zeta) = \int_{\eta'} \frac{b}{(g-\zeta)^{k+1}}.$$

Divide this integral into two pieces according to $\eta' = \eta'' + (g^*v)F_{T_2}$. The distance from the pie-slice T to $Supp_{\mathbf{C}}(\eta'')$ is bounded below, so $|(g(z)-\zeta)^{-k-1}| \leq C^k \leq C^n$ for z in the support of η'' and ζ in T. On the other hand, η'' satisfies an estimate of the form $\mathbf{F}(\eta''_n, \varepsilon n, C, \varepsilon)$ as described at the corresponding place in the first half of this section. The integrand b_I vanishes if I has any repetitions, so putting these together we find

$$\left|\int_{\eta''_n} \frac{b}{(g-\zeta)^{k+1}}\right| \leq C^n n^{-\varepsilon n}.$$

The chain $(g^*v)F_{T_2}$ is supported in $\Re x \geq \xi_0 - 4\epsilon - L$, and it satisfies the estimate $\mathbf{E}$ given in Corollary 8.2. If $\zeta \in T$ and $z \in Supp_{Z_*}(F_{T_2})$, then

$$|(g(z)-\zeta)^{k+1}| \geq (\xi_1 - \Re g(z))^{k+1},$$

which is uniform in ζ. Therefore

$$\begin{aligned}\left|\int_{(g^*v)(F_{T_2})_n} \frac{b}{(g-\zeta)^{k+1}}\right| &\leq C^n \int_{x=0}^{L/2+4\epsilon} x^{-k-1} \Re g_* \|F_{T_{2,n}}\|(\xi_1 - x) \\ &\leq C^n n^{-\varepsilon n} \int_{x=0}^{L} x^{-k-1} \omega(\varepsilon n)(-x),\end{aligned}$$

the last line by the estimate 8.2. Recall that $\omega(\varepsilon n)(-x) = x^{\varepsilon n-1}dx$. We have

$$\int_0^L x^{-k-1}\omega(\varepsilon n)(-x) = \int_0^L x^{-k+\varepsilon n-2}dx \leq C^n,$$

uniformly for all n such that $\varepsilon n - 2 \geq k$. Therefore

$$\left|\int_{(g^*v)(F_{T_2})_n} \frac{b}{(g-\zeta)^{k+1}}\right| \leq C^n n^{-\varepsilon n}.$$

This estimate is uniform for $\zeta \in T$. Putting this together with the integral over η'' we get

$$\left| \int_{\eta'_n} \frac{b}{(g-\zeta)^{k+1}} \right| \le C^n n^{-\varepsilon n},$$

so

$$\left| \frac{1}{k!} \frac{d^k}{d\zeta^k} f_n(\zeta) \right| \le C^n n^{-\varepsilon n}.$$

This proves condition (2.5.3).

9. BOUNDS ON MULTIPLICITIES

In this section we will use the assumption that the chain η is obtained from β by finitely many applications of the procedure outlined in §§4–8, to prove that if z is a generic point of $\Lambda(\ell) = \Lambda(\ell_1) \times \ldots \times \Lambda(\ell_n)$ then the multiplicity of $FH\varphi$ at z is bounded by C^n. This was an ingredient in the previous section's proof of (2.5.3).

First we give a general description of the chains which can arise from repeated applications of the procedures outlined in the previous section.

Suppose A and B are subsets of $\mathbf{R}^a$ and $\mathbf{R}^b$ respectively. A continuous map $f : A \to B$ is *piecewise polynomial* if there is a finite decomposition $A = \bigcup U_\alpha$ and if there are polynomial maps P_α from $\mathbf{R}^a$ to $\mathbf{R}^b$ such that $f|_{U_\alpha} = P_\alpha$. It follows from continuity that the boundaries between pieces are algebraically defined. The *degree* of f is the largest of the degrees of the component polynomials $P_{\alpha,i}$, $i = 1, \ldots, b$. If s_i and t_j are coordinates in $\mathbf{R}^a$ and $\mathbf{R}^b$ respectively, then the *degree of f_j in the variable s_i* is the largest of the degrees of the polynomials $P_{\alpha,j}$ in the variable s_i. The *size* of f is the largest of the following numbers: the number of pieces U_α, and the suprema of the derivatives $\sup_{U_\alpha} |\partial P_{\alpha,i}/\partial x_j|$.

If $f : A \to B$ and $g : B \to C$ are piecewise polynomial maps, then $g \circ f : A \to C$ is a piecewise polynomial map. Furthermore, the degree of $g \circ f$ is less than or equal to the product of the degrees of f and g. The number of pieces into which the map $g \circ f$ is decomposed is less than or equal to the product of the number of pieces for g and the number of pieces for f. The supremum of the partial derivatives of $g \circ f$ is less than or equal to $dim(B)$ times the product of the suprema of the partial derivatives of f and g, by the chain rule.

We now define a type of piecewise polynomial map which will arise in describing the chains that can occur. Suppose we have a graph organized as a tree, with some edges marked and some not, beginning with m vertices along the top and ending with n vertices along the bottom. Suppose that for each marked edge we have a map $f(e, x, t) : Z \times [0, 1] \to Z$, and suppose that for each unmarked edge e we have a map $f(e, x) : Z \to Z$. Suppose that there are N marked edges. Then we get a map

$$\Phi : Z^m \times [0, 1]^N \to Z^n$$

defined as follows (it is similar to our usual construction seen first in §5). Fix $z \in Z^m$ and $s \in [0, 1]^N$. For each vertex v in the graph we will get a point $\phi(v) \in Z$. If $v_1, \ldots, v_n$ are the vertices at the bottom, we will set

$\Phi(z,s) = (\phi(v_1),\ldots,\phi(v_n))$. Define the points $\phi(v)$ inductively as follows. Let e be the edge above a vertex v, and let u be the vertex above e. If e is a marked edge, let s_i denote the corresponding coordinate. In that case, $\phi(v) = f(e,\phi(u),s_i)$. If e is an unmarked edge then $\phi(v) = f(e,\phi(u))$.

We will generally suppose that a subset $U \subset Z^m \times [0,1]^N$ is fixed, and require that there is a relatively compact open set $Z' \subset Z$ such that in defining $\Phi|_U$, all of the intermediate points $\phi(v)$ lie in Z'. Suppose that each $f(e,x)$ or $f(e,x,t)$, when restricted to Z' or $Z' \times [0,1]$, is piecewise polynomial of degree and size less than C_0. Suppose that there are M edges in the graph altogether. There is C_1 depending only on C_0 such that Φ is piecewise polynomial of size less than C_1^M. Suppose e is a marked edge with associated parameter s_i and suppose v_j is a bottom vertex. Then the degree of $\Phi_j|_U$ in the variable s_i is less than C_1^d where d is the number of edges from e (inclusive) to v. If v is not below e then Φ_j doesn't depend on s_i so the degree is zero.

If $\Phi : Z^m \times [0,1]^N \to Z^n$ and $\Psi : Z^n \times [0,1]^P \to Z^p$ are maps derived from graphs in the above way, then the composition $\Psi \circ \Phi : Z^m \times [0,1]^{N+P} \to Z^p$ is derived from the union of the two graphs, where the n bottom vertices in the graph associated to Φ are joined with the n top vertices in the graph associated to Ψ.

We now describe a condition which will be satisfied by the graphs associated to the maps we need below. It is essentially the condition described in Proposition 5.2. Let r be some number. We say that the marked edges in a graph are *equidistributed with r exceptions* if there are r triple branch points in the graph and a map from the rest of the triple branch points to the set of marked edges, which is injective and which maps each branch point to an adjoining edge.

Proposition 9.1 *Fix finitely many applications of the procedures of §§4–8, including a number M such that all points which occur are beyond points of β_* at distance $\leq M$. Any chain η on Z_I ($|I| = n$) which arises as a result of the procedures is a sum of C^n components each of which has the following form. (We will denote the component also by η.) It is an m-chain, where $m - n$ is bounded independent of I. There are numbers a, b, and r, with $b = a - m$, and b and r bounded independent of I. There is a map $p : [0,1]^a \to Z_I$ and there are maps $q_i : [0,1]^a \to Z_{I^i}$, for $i = 1,\ldots,b$, and differential forms ω_i on* **C**. *The chain η is given as follows. Suppose u is an m-form on Z_I. Then*

$$\int_\eta u = \int_{[0,1]^a} p^*(u) \wedge q_1^*(g^*\omega_1) \wedge \ldots \wedge q_b^*(g^*\omega_b).$$

Let $U \subset [0,1]^a$ denote the inverse image $p^{-1}(\overline{Z}_I)$. The maps $p|_U$ and $q_i|_U$ are obtained from maps of the form $\Phi : Z \times [0,1]^a \to Z^n$ which come from graphs with one top vertex as described above, by putting the point P into the Z coordinate. The flows and maps $f(e)$ which occur in the definition of the maps Φ are of the form $f(e,x,t) = f_{ij}(x,t)$ or $f(e,x) = f_{ij}(x,t_0)$ with t_0 fixed; where f_{ij} denotes a flow which occurs in some application of the process of §§4-8. In the graphs used to define the maps, the marked edges are equidistributed with $\leq r$ exceptions. There are less than Cn edges in any graph which occurs. There are an M and a subset $U \subset [0,1]^a$ such that the points in $p(U)$ or $q_i(U)$ are a priori *beyond points of β_* at distance $\leq M$. If R is the radius given by Proposition 6.1, then Z' be the set where $d(z,P) \leq R$. The intermediate points $\phi(v)$ in the constructions $\Phi|_U$ all lie in Z'. The form $q_1^*(\omega_1) \wedge \ldots \wedge q_b^*(\omega_b)$ is supported on U. The maps $p|_U$ and $q_i|_U$ are piecewise polynomial of size less than C^n.*

Proof: The chain η is a component of a chain Y' obtained from β by finitely many of the following types of operations.
(1) Applying finitely many of the operators $(1-AK)^{-1}$, $(1+AK)^{-1}$, H, A, or K, and then applying the operator F (with respect to some choice of flows).
(2) Multiplying by $g^*(\omega)$ for some differential form ω on $\mathbf{C}$.
We have to prove that these operations preserve the above form of the chains. It is clear that β itself has the requisite form (one can define a piecewise linear map from the cube $[0,1]^n$ to the n-simplex, in the form of a map corresponding to a graph as required).

The first type of operation preserves the form of the chain, because the description given in §7 implies that there is a map $\Phi : Z^n \times [0,1]^u \to Z^{n'}$ which comes from a graph as described above, such that $\eta' = \Phi(\eta \times [0,1])$. Then to get the new map p' just take $p' = \Phi \circ p$; we have $a' = a + u$; and keep the same maps q_i and forms ω_i. The graph associated to Φ is equidistributed with finitely many exceptions by Proposition 5.2, and it has $2n' - n$ edges, so the required properties of the graphs are preserved when this one is added onto the graph corresponding to p.

It is also easy to see that the second type of operation preserves the form of the currents—one just adds the map p to the list of q_i and the form ω to the list of ω_i.

Fix a number M, so that in the finite collection of procedures, the eventually cut off chains are always supported on sets of points beyond β_* at distance $\leq M$. We can let $U \subset [0,1]^a$ be the set of points t such that the resulting $\Phi(t)$ are *a priori* beyond points of β_* at distance $\leq M$. Just count up the distance travelled by the various flows except for the f^0. The form on $[0,1]^a$ which defines η will be supported on U.

Let R be the radius given by Proposition 6.1, depending on M. Let Z' denote the relatively compact open set $d(x,P) < R$ inside Z. When defining $\Phi(t)$ for $t \in U$, the intermediate points $\phi(v)$ will be contained in Z'. Let C_0 be an upper bound for the degrees and sizes of the flows f_{ij} in the region Z'. The number of edges in the graph defining Φ is bounded by a multiple of n. As noted above, there is C such that $\Phi|_U$ is piecewise polynomial of size less than C^n. This completes the proof of the proposition.

Remark: For each marked edge e_i, with corresponding variable s_i, and each bottom vertex v_j below e_i, let δ_{ij} be the number of edges in the graph defining p, from e_i (inclusive) to v_j. The degree of $p_j|_U$ in the variable s_i is less than $C^{\delta_{ij}}$. If v_j is not below e_i then the degree of p_j is the variable s_i is zero.

Lemma 9.2 *Suppose $f : [0,1]^n \to \mathbf{R}^{n-r}$ is a polynomial map. Let d_{ij} denote the degree of f_j in the variable s_i. Let*

$$D = \max_{\substack{S\subset[1,n] \\ |S|=n-r}} \sum_{\sigma:S\hookrightarrow[1,n-r]} \left(\prod_{i\in S} d_{i\sigma(i)}\right).$$

Suppose that the fiber $f^{-1}(0)$ is generically a smooth complete intersection. Then the volume of $f^{-1}(0)$ is less than $C^n D$.

Proof: Let $V \subset f^{-1}(0)$ be the open subset where the fiber is a smooth complete intersection. It has full measure so we must find its volume. For each projection $p : [0,1]^n \to [0,1]^r$, let V_p denote the open set of points u where the Jacobian $|\wedge^r dp(u)|$ is larger than the Jacobians $|\wedge^r dq(u)|$ for all other projections. The union $\bigcup_p V_p$ is an open set of full measure, so it suffices to calculate its volume. Note that there are $\binom{n}{r} \leq C^n$ projections p so it suffices to compute the volume of a single V_p. Since V_p and $[0,1]^r$ have the same dimension, we have the following estimate:

$$vol(V_p) \leq \frac{deg(p|_{V_p})}{min_{V_p}|\wedge^r dp|}$$

where $deg(p)$ denotes the maximum of the degrees of the covering $p : V_p \to [0,1]^r$ at generic points. We have

$$min_{V_p} |\wedge^r dp| \geq \binom{n}{r}^{-1/2}.$$

To prove this, note that for $u \in V_p$, $|\wedge^r dp(u)| = \sup_q |\wedge^r dq(u)|$ where the supremum is over the $\binom{n}{r}$ projections q. But these projections provide an orthonormal basis for the dual space of $\bigwedge^r \mathbf{R}^n$, so if $w \in \bigwedge^r TV$ is an r-vector, then $|w|^2 = \sum_q |dq_* w|^2$. Apply this to an r-vector w of norm 1, and note that $|dq_* w| = |\wedge^r dq|$. In particular, $|q_* w| \leq |\wedge^r dp(u)|$. Thus we have

$$1 \leq \binom{n}{r} |\wedge^r dp(u)|^2$$

as desired. Now $\binom{n}{r} \leq 2^n$ so putting this together with the previous estimate, we get

$$vol(V_p) \leq C^n deg(p|_{V_p}).$$

Now let us compute $deg(p|_V)$. Let $S \subset [1,n]$ denote the complement of the set of coordinates used in the projection p. Thus $|S| = n - r$. Let $s \in [0,1]^r$ be a generic point. Restricting to the $[0,1]^{n-r}$ which is the fiber of p over s, we have a map $f : [0,1]^{n-r} \to \mathbf{R}^{n-r}$ and $f^{-1}(0)$ is a finite set. The number of elements of this set is the degree of the projection p over the point s. We claim that the number of elements is $\leq D$.

In general suppose that we have a map $f : \mathbf{C}^m \to \mathbf{C}^m$ such that the degree of f_j in the variable s_i is d_{ij}. Suppose that $f^{-1}(0)$ is a finite set. Let x_j be the coordinate functions on the range. Then $u_j = f^*(x_j)$ are polynomials on $\mathbf{C}^m$ which have degrees d_{ij} in the variables s_i respectively. We have to show that if the number of points in the locus of common zeros of the u_j is finite, it is less than

$$D_1 = \sum_{\sigma:[1,m]\hookrightarrow[1,m]} \left(\prod_{i=1}^{m} d_{i\sigma(i)} \right).$$

Compactify $\mathbf{C}^m$ to $(\mathbf{P}^1)^m$. Let L_i denote the line bundles which are pullbacks of $\mathcal{O}_{\mathbf{P}^1}(1)$ by the coordinate projections. Think of u_j as sections of the line bundles

$$M_j = \bigotimes_{i=1}^{m} L_i^{\otimes d_{ij}}.$$

The u_j have only finitely many common zeros in $\mathbf{C}^m$ but may have components of higher dimension in their common zeros at infinity. However, we contend that this doesn't happen generically, in other words that for generic sections v_j of M_j, $j = 1, \ldots, m$, the intersection $Z(v_1) \cap \ldots \cap Z(v_m)$ is a finite set. Assume for the moment that this is the case. Then the number of points in the intersection, counted with multiplicity, is given by intersection theory, and is equal to

$$c_1(M_1)c_1(M_2)\ldots c_1(M_m).$$

Let $H_i = c_1(L_i)$. The intersection number is the intersection of the classes $\sum d_{ij}H_i$ for $j = 1, \ldots, m$. We have $H_iH_i = 0$ and $H_1H_2 \ldots H_m = 1$. Therefore the intersection of the classes $\sum d_{ij}H_i$ is equal to

$$D_1 = \sum_{\sigma:[1,m] \hookrightarrow [1,m]} \left(\prod_{i=1}^{m} d_{i\sigma(i)} \right).$$

Now the actual number of closed points in the intersection of the $Z(v_j)$ is less than the sum of the points counted with multiplicities (all multiplicities are positive), so the number of points is less than D_1. On the other hand if the sections u_j are moved slightly to become generic, then the finitely many points in their common intersection in $\mathbf{C}^m$ will move slightly, and possibly break into more points, but in any case the number of points in the common intersection of the generic sections will be greater than or equal to the number of points in the intersection of the u_j. Thus the number of points in the intersection of the u_j is less than D_1 so this will prove the lemma. Therefore we only have to prove the contention that generic sections v_j intersect properly.

First we show that the linear systems $H^0(M_j)$ have no base locus. Indeed, the line bundles L_i on $(\mathbf{P}^1)^m$ are invariant under any automorphism of a single factor or any composition of such. Therefore the same is true for the M_j. Thus if there is a base locus, it is invariant under translations in any factor. But these translations form a transitive group action, so if there is a base locus, it would be equal to the whole space. However, since we already have nonzero sections u_j of M_j, this latter is not the case. Therefore there is no base locus. Now in general if a linear system of divisors has no base locus, and if Y is any subvariety, then the restriction of a generic divisor in the linear system to Y is a divisor on Y. If we apply this inductively to $Y_j = Z(v_1) \cap \ldots \cap Z(v_j)$ we conclude that Y_{j+1} is a divisor on Y_j. In particular Y_m is a finite set. This completes the proof of the lemma.

The following proposition fills in the gap in the argument of §8 (Proposition 8.1).

Proposition 9.3 *Suppose $\ell = (\ell_1, \ldots, \ell_n)$ is a critical point on Z_I, with $|I| = n$. Suppose z is a generic point of $\Lambda(\ell) = \Lambda(\ell_1) \times \ldots \times \Lambda(\ell_n)$. Then the multiplicity of the chain $FH\varphi$ at the point z is less than C^n.*

Proof: The chain $FH\varphi_I$ is a sum of C^n chains of the form given by Proposition 9.1, so we may restrict our attention to one such. $FH\varphi_I$ is an n-chain, so $a = n + b$. Note that b is given by the number of procedures, so it is bounded independently of n. There is an open set $U' \subset U \subset [0,1]^a$ such that p maps U' to $\Lambda(\ell) \subset Z_I$. We need only consider the current restricted to U'. For a generic point $z \in \Lambda(\ell)$, the fiber $p^{-1}(z)$ is piecewise smooth of real codimension n in U'. The multiplicity in question is the integral

$$\int_{p^{-1}(z)} q_1^*(g^*\omega_1) \wedge \ldots \wedge q_r^*(g^*\omega_r).$$

On U' the maps q_i are piecewise polynomial of sizes less than C^n so we get bounds $\sup |q_i^*(g^*\omega_i)| \leq C^n$. Thus it suffices to prove that the b dimensional volume of $p^{-1}(z)$ is bounded by C^n. The fiber $p^{-1}(z)$ decomposes into pieces corresponding to the different parts of the piecewise polynomial map. There are less than C^n pieces, so we may restrict our attention to one. We will use Lemma 9.2 to bound the volume. Recall the graph with flows and marked edges used to construct the map p. The remark after 9.1 says that for each marked edge e_i, with corresponding variable s_i, and each bottom vertex v_j below e_i, the degree of p_j in the variable s_i is less than $C^{\delta_{ij}}$, where δ_{ij} is the number of edges from e_i (inclusive) to v_j. If v_j is not below e_i then the degree of p_j is the variable s_i is zero. Thus by Lemma 9.2, the volume of $p^{-1}(z)$ is less than $C^n D$ where

$$D = \max_{\substack{S \subset [1,a] \\ |S| = n}} \sum_{\sigma : S \hookrightarrow [1,n]} \left(C^{\sum_{i \in S} \delta_{i\sigma(i)}} \right),$$

the first sum being taken over the maps σ such that $v_{\sigma(i)}$ is below e_i. We will show that $D \leq C^n$ by fixing an $S \subset [1,a]$ and showing that the number of maps $\sigma : S \hookrightarrow [1,n]$ such that $v_{\sigma(i)}$ are below e_i is less than C^n, and that given such a map the sum $\sum_{i \in S} \delta_{i\sigma(i)}$ is less than Cn. Once we have fixed our subset S, we may consider as marked only those edges e_i which correspond to $i \in S$. The map p restricted to the $[0,1]^n$ determined by S is given by the same construction as before, but with only the marked edges described above; the other edges revert to being unmarked. For each edge e of the graph, let $A(e)$ denote the difference, the number of bottom vertices below e minus the number of marked edges (in the present sense) strictly below e. Choose our maps σ from the bottom up, in other words at each stage, fix $\sigma(i)$ where e_i is a minimum in the obvious partial ordering of edges. The number of choices for $\sigma(i)$ is equal to the number of bottom edges below e_i, minus the number which are already used up, in other words the number of marked edges strictly below e_i. Thus the number of choices for $\sigma(i)$ is less than or equal to $A(e_i)$. Now $A(e_i) \leq C^{A(e_i)}$ so the number of choices of σ is less than

$$C^{\sum_{e_i} A(e_i)}.$$

Thus it suffices to show that $\sum_e A(e) \leq Cn$. On the other hand, suppose a map σ is given. Then we claim that there is also a bound for the sum

$$\sum_{e_i} \delta_{i\sigma(i)} \leq \sum_{e} A(e).$$

To see this, add up for each edge e the number of paths from e_i to $v_{\sigma(i)}$ which go through e. Then add these for all edges. Given e, no more than $A(e)$ paths can go through e. This is because the number of paths is less than the number of bottom vertices below e which are left after taking away the $v_{\sigma(i)}$ for all edges e_i below e. Therefore the sum for all edges is less than $\sum_e A(e)$ as required. So for the second task it also suffices to prove that $\sum_e A(e)$ is less than or equal to Cn.

To prove $\sum_e A(e) \leq Cn$, we use the equidistribution condition. Note that after marking only the edges corresponding to S, the marked edges in the graph are still equidistributed, this time with $\leq r'$ exceptions, for some r' independent of n. Therefore if e is an edge, then the number of marked edges strictly below e is at least equal to the number of forks below e, minus $1 + r'$. But the number of bottom edges below e is equal to the number of forks minus 1. Therefore $A(e) \leq r'$. On the other hand, the total number of edges in the graph is less than Cn. Therefore $\sum_e A(e) \leq Cn$. This completes the proof of the proposition.

10. Regularity of individual terms

We have now shown that the individual terms in the Laplace transform add up to the Laplace transform of the monodromy. Finally we have to show that the individual terms have regular singularities. This will verify properties *(2.5.4)* and *(2.5.5)* from §2. The formal sum of the power series for the individual terms will then give the power series for the singularities of the Laplace transform of the monodromy, due to the estimates given in the previous section.

Condition (2.5.4)

First we must prove that the terms $f_n(\zeta)$ have locally finite regular singularities. To do this we use the following proposition, a technical extension of the well-known regularity of the Gauss-Manin connection.

Suppose (Y, Y') is a pair consisting of a complex manifold and a closed analytic subset. Suppose $g : Y \to \mathbf{C}$ is a holomorphic function, and b is a holomorphic form of top degree. Suppose that for each ζ in the universal cover of $D^*(s, \epsilon)$, we have a cycle in relative homology $\eta(\zeta)$, such that ζ is not contained in the support of $g_*\eta(\zeta)$, and such that for ζ' near ζ, $\eta(\zeta')$ is homologous to $\eta(\zeta)$ by a homology $\kappa(\zeta, \zeta')$ whose support doesn't meet ζ (in other words, $\eta(\zeta') - \eta(\zeta) = \partial\kappa(\zeta, \zeta')$ in relative homolory). Then the function

$$f(\zeta) = \int_{\eta(\zeta)} \frac{b}{g - \zeta}$$

is a multivalued analytic function on $D^*(s, \epsilon)$.

Proposition 10.1 *Suppose that Y' is a divisor with normal crosings in Y, and suppose that the critical point sets of the function g on (Y, Y') are compact. Suppose there is a compact subset $K \subset Y$ such that for any ζ in the universal cover of $D^*(s, \epsilon)$, the cycle $\eta(\zeta)$ is contained in K. Suppose also that the homologies $\kappa(\zeta, \zeta')$ are contained in K. Then the function $f(\zeta)$ has regular singularities.*

Proof: Since the critical point sets are compact, we may make a resolution of singularities $\tilde{Y} \to Y$, such that the fibers of g are divisors with normal crossings and the strict transform $\tilde{Y}'$ of Y' is a divisor with normal crossings, which crosses fibers of g normally [12].

Without loss of generality, we may assume that $s = 0$ is at the origin in the complex plane. Let Y_0 denote the fiber over the origin. Following Clemens

[5], there is a neighborhood $\tilde{U}$ of the compact set $\tilde{K} \cap \tilde{Y}_0$, a retraction π from $(\tilde{U}, \tilde{U}')$ to $(\tilde{U}_0, \tilde{U}'_0)$, and an action of S^1 on $(\tilde{U}, \tilde{U}')$ which commutes with π and covers (via g) the rotation action on the complex plane. This provides an action of the monodromy transformation T on the space of relative cochains in a fiber $(\tilde{U}_\zeta, \tilde{U}'_\zeta)$, and an action on the Leray spectral sequence for the projection

$$\pi : (\tilde{U}_\zeta, \tilde{U}'_\zeta) \to (\tilde{U}_0, \tilde{U}'_0).$$

Let $\mathcal{F}$ denote the sheaf which calculates relative cohomology, $H^i(\tilde{U}_\zeta, \mathcal{F}) = H^i(\tilde{U}_\zeta, \tilde{U}'_\zeta)$. As described in the papers of Clemens [5] and Landman [16], and Griffiths's appendix to Landman's paper, a power of the monodromy T^N acts trivially on the sheaves $R^q\pi_*\mathcal{F}$, and hence on the $H^p(\tilde{U}_0, R^q\pi_*\mathcal{F})$ term of the spectral sequence. Since the spectral sequence converges to the associated graded of a filtration on the relative cohomology, T^N acts unipotently: $(T^N - I)^K = 0$ on $H^*(\tilde{U}_\zeta, \tilde{U}'_\zeta)$. The same is then true for relative homology.

There is a small disc $D(0, \epsilon)$ so that $g^{-1}D(0, \epsilon) \cap \tilde{K} \subset \tilde{U}$. If ζ is in this disc, then by piecing together homologies $\kappa(\zeta, \zeta')$ we can form a homology $\kappa(\zeta, T\zeta)$ between $\eta(\zeta)$ and $\eta(T\zeta)$. In other words $\partial\kappa(\zeta, T\zeta) = \eta(T\zeta) - \eta(\zeta)$, so

$$(T - I)f(\zeta) = \int_{\partial\kappa(\zeta, T\zeta)} \frac{b}{g - \zeta}.$$

Let $\mu(\zeta) = \kappa(\zeta, T\zeta) \cap Y_\zeta$. Find a section c of $\Omega^{n-1}_{Y/\mathbf{C}}$ so that $c \wedge dg = b$. It is regular where g is smooth. By the residue theorem,

$$\int_{\partial\kappa(\zeta, T\zeta)} \frac{b}{g - \zeta} = \int_{\mu(\zeta)} c.$$

Now $\kappa(\zeta, T\zeta) \subset K$, so $\mu(\zeta)$ represents a class in $H_{n-1}(\tilde{U}_\zeta, \tilde{U}'_\zeta)$. As ζ moves, the cycles are homologous to each other by homologies relating the various fibers of g. By the preious paragraph, $(T^N - I)^K \mu(\zeta) = 0$. Therefore $(T^N - I)^K (T - I)f(\zeta) = 0$.

By using Clemens' retraction, one can deform $\eta(\zeta)$ to a cycle of integration whose support in $\mathbf{C}$ misses a sector going to the origin (although the support will still contain the origin). This cycle has finite volume, so the integral of $(g - \zeta)^{-1}b$ will be bounded by $C|\zeta|^{-1}$. By the criterion of Lemma 2.2, $f(\zeta)$ has a regular singularity at the origin.

Remark: The number N in the quasi-unipotence of f is the least common multiple of the multiplicities of divisors which occur in the resolutions of the singularities in the compact subset K.

We now apply this to our situation. We would like to express the integrals $f_I(\zeta)$ as integrals in a space where the critical point sets of g are compact. This will take some work, making use of the assumption that our original Riemann surface S was compact.

For each index I, let Γ_I denote the subgroup of translations of Z_I by elements of $\pi_1(S)^n$ which preserves the function g. Then b_I and g descend to the quotient Z_I/Γ_I. Furthermore, if $\alpha : I' \to I$ is an elementary arrow then the map $\alpha : Z_{I'} \to Z_I$ descends to a map $Z_{I'}/\Gamma_{I'} \to Z_I/\Gamma_I$. To see this, suppose $\gamma' = (\gamma_1', \ldots, \gamma_m') \in \Gamma_{I'}$. We will find $\gamma \in \Gamma_I$ such that $\alpha \circ \gamma' = \gamma \circ \alpha$. Suppose α is determined by a number l as in §4. Set $\gamma_k = \gamma_k'$ for $k \leq l$ and $\gamma_{k+1} = \gamma_k'$ for $k \geq l$. Then $\alpha(\gamma' z)_k = \gamma_k' z_k = \gamma_k \alpha(z)_k$ if $k \leq l$ and $\alpha(\gamma' z)_{k+1} = \gamma_k' z_k = \gamma_{k+1}\alpha(z)_{k+1}$ if $k \geq l$, so $\alpha \circ \gamma' = \gamma \circ \alpha$. Therefore the integrals can be defined as integrals over relative homology classes on the spaces Z_I/Γ_I. We have to show that the connected components of the critical point sets are compact.

Fix I with $|I| = n$. The critical point sets on the relative space Z_I/Γ_I are the unions of some subsets, each of which is the image under a composition of elementary arrows $J \to I$ of the critical point set of g_J on the absolute—not relative—space Z_J/Γ_J. We will call the image of a component of the critical point set of g_J on the absolute Z_J/Γ_J a *subsidiary*. Each component of the full critical point set is the union of a collection of subsidiaries, minimal with respect to the condition that any other subsidiary which meets a member of the collection should also be included. We will prove compactness of the component in two steps, first proving that each subsidiary is compact, and then proving that there are only finitely many subsidiaries in each component.

Write $Z_J = Z^a \times Z^b$, where the a coordinates are those for which $j_{k-1} \neq j_k$ and the b coordinates are those for which $j_{k-1} = j_k$. The function g is constant in the Z^b direction, in other words it is pulled back from a function g' on Z^a. The function g' is a sum of functions on the factors, each of which has isolated critical points. The critical points on Z^a are those for which each coordinate is a critical point for the appropriate summand of g', in other words the critical points of g' on Z^a are isolated. Since the function g is constant in the factor Z^b, the critical point set of g on $Z^a \times Z^b$ is the product of the critical point set of g' on Z^a, with Z^b. Thus the components of the critical point set on $Z^a \times Z^b$ are of the form $\{z'\} \times Z^b$. On the other hand, the group $\Gamma = \Gamma_J$ contains all of the elements of $\pi_1(S)^b$, since the function g is constant in the Z^b direction. Thus $\Gamma = \Gamma' \times \pi_1(S)^b$ where Γ' is the subgroup of $\pi_1(S)^a$ which preserves the function g' on Z^a. Therefore $Z_J/\Gamma_J = (Z^a/\Gamma') \times S^b$. The pieces of the critical point set for g are of the form $\{z'\} \times S^b$ where z' is a critical point of g' on Z^a/Γ'. In particular, since S is compact, these pieces are compact. The images of these pieces in Z_I/Γ_I are the subsidiaries of the critical point set of g, hence they are compact.

In order to prove that there are only finitely many subsidiaries in any component, we must give a combinatorial description of the subsidiaries. First of all, say that a *roster* W is a directed row of edges connecting vertices, an assignment of an index $j(v)$ for each vertex, and for each edge e such that $l(e) \neq r(e)$, an assignment of a critical point $w(e)$ of $g_{l(e)r(e)}$ on Z. Here $l(e) = j(v_1)$ and $r(e) = j(v_2)$ where v_1 and v_2 are the vertices on the left and right ends of the edge e respectively. To every roster there corresponds an index J: number the vertices in the row $v_0, \ldots, v_m$ from left to right and put $j_k = j(v_k)$. We have $|J| = m$. Furthermore, to a roster there corresponds a piece of the critical point set of the function g_J on the (absolute) space Z_J, obtained as follows. Number the edges in the roster $e_1, \ldots, e_m$ from left to right. Then the corresponding piece of the critical point set is the set of points $z = (z_1, \ldots, z_m)$ such that $z_k = w(e_k)$ if $j_{k-1} \neq j_k$ (z_k is arbitrary if $j_{k-1} = j_k$).

There is an action of the group Γ_J on the set of rosters with index J, obtained as follows. If $\gamma = (\gamma_1, \ldots, \gamma_m)$ and if W is a roster, then $W' = \gamma W$ is a roster with the same indices but with different critical points: $w'(e_k) = \gamma_k w(e_k)$ whenever $j_{k-1} \neq j_k$. This action commutes with the correspondence between rosters with index J and components of the critical point set of g_J on Z_J. Therefore the correspondence descends to a correspondence between rosters up to the action of Γ_J, and pieces of the critical point set of g on the absolute Z_J/Γ_J.

A subsidiary is the image of a component of the critical point set of g on Z_J/Γ_J, under a map $\alpha : J \to I$. Therefore a subsidiary corresponds to a pair (W, α) where W is a roster with index J and $\alpha : J \to I$, up to the action of the group Γ_J on the roster W.

We will define an invariant of a roster, then show that up to Γ_J, there are only finitely many rosters having that invariant. Finally we will show that if two subsidiaries intersect, their rosters have the same invariant. Since there are only finitely many possibilities for the map α, this will complete the proof of the finiteness and hence of the compactness.

The invariant of a roster will be another roster U together with a complex number μ, with the following properties, and up to the following group action. First, its index K has no repetitions, in other words $k_{i-1} \neq k_i$ for all i. In particular, there are critical points $u(e)$ for every edge. The second property is that there are no repetitions in the images of the critical points in S, in other words $p(u(e_{i-1})) \neq p(u(e_i))$ for all i, where $p : Z \to S$ is the projection. The group action is an action of $\pi_1(S)^r$, where r is the length of the roster. The action of $\pi_1(S)^r$ on rosters of length r is the same as above, an element $(\gamma_1, \ldots, \gamma_r)$ sending a roster U to U' with the same index but with critical points $u'(e_i) = \gamma_i u(e_i)$. There is a map $\nu_J : \pi_1(S)^r \to \mathbf{C}$ such that if $z \in Z_J$ then $g_J(\gamma z) = g_J(z) + \nu_J(\gamma)$. (Hence Γ_J is the kernel of ν_J.) The action of $\pi_1(S)^r$ on pairs (U, μ) is given by $\gamma(U, \mu) = (\gamma U, \mu + \nu_J(\gamma))$ where J is the index of U. We consider invariants (U, μ) up to the action of $\pi_1(S)^r$.

We will define an invariant (U, μ) of pairs (W, ϖ) (where W is a roster and ϖ is a complex number). Then for a roster W, define its invariant to be the invariant of $(W, 0)$. Given a pair (W, ϖ), obtain its invariant in a sequence of steps of the following form. Begin with $U = W$ and $mu = \varpi$. One step is to contract an edge e with $l(e) = r(e)$, and replace it with a single vertex, assigned the index $l(e) = r(e)$. The other step can be done if there are two consecutive edges e and e' such that $l(e) \neq r(e)$ and $l(e') \neq r(e')$, and such that $pu(e) = pu(e')$ in S. Act on (U, μ) by the element $(1, \ldots, \gamma, \ldots, 1) \in \pi_1(S)^r$ with γ in the place corresponding to e', so that $u(e) = \gamma u(e')$ in Z. Then remove the vertex between the two edges, joining the two edges to become one, and assign the point $u(e) = \gamma u(e')$ to this new edge (if a point need be assigned). To obtain the invariant, repeat these two steps until a pair (U, μ) with the desired properties is reached. We remark that if (W, ϖ) and (W', ϖ') are equivalent by $\pi_1(S)^r$, then they give equivalent invariants. Furthermore, the invariant does not depend on the sequence of steps used to reach it. To see this, it suffices to show that if (U, μ) and (U', μ') are obtained from (W, ϖ) by different steps, then there is (V, ν) which can be obtained from either (U, μ) or (U', μ') in one step. If the steps used to obtain U and U' involve different edges, then this is clear. The other possibility is that there are three edges e, e' and e'', and that e and e' are joined to obtain U, but e' and e'' are joined to obtain U'. In either case, the newly formed edge can be joined to the third edge, to get V. Thus the invariant (U, μ) of a pair (W, ϖ) is well defined.

We claim that up to the action of $\pi_1(S)^r$, there are only finitely many pairs (W, ϖ) which give the invariant (U, μ). This will prove that there are only finitely many rosters W which give invariant (U, μ), up to the action of Γ_J, because Γ_J is the kernel of the action on the second variable.

In fact, we claim that if (U, μ) is a pair, then there are only finitely many pairs (W, ϖ) up to equivalence which yield a pair equivalent to (U, μ) after one of the operations described above. Since $n = |I|$ is fixed, there is a bound on the number of operations, so this will prove the desired finiteness.

If the operation involves contracting an edge e with $l(e) = r(e)$, then the pair (U, μ) determines (W, ϖ) by $w(e') = u(e')$ for $e' \neq e$. Equivalent pairs (U, μ) determine equivalent pairs (W, ϖ).

Suppose the operation involves joining two consecutive edges e and e' together, with $l(e) \neq r(e')$. Up to equivalence, we may assume that $w(e) = w(e')$. Then $w(e)$ and $w(e')$ are equal to $u(e'')$ where e'' denotes the resulting edge. Again, equivalent pairs (U, μ) yield equivalent pairs (W, μ) because if $u(e'')$ is replaced by $\gamma u(e'')$ then $w(e)$ and $w(e')$ may be replaced by $\gamma w(e)$ and $\gamma w(e')$. Note that the term $\nu_I(1, \ldots, \gamma, \gamma, \ldots, 1)$ is equal to $\nu_{J'}(1, \ldots, \gamma, \ldots, 1)$ if J' and J are the indices associated to W and U respectively. This is because

$$g_{l(e)r(e)}(\gamma z) - g_{l(e)r(e)}(z) + g_{l(e')r(e')}(\gamma z) - g_{l(e')r(e')}(z) = g_{l(e'')r(e'')}(\gamma z) - g_{l(e)r(e')}(z)$$

as $l(e'') = l(e)$, $r(e'') = r(e')$, and $l(e') = r(e)$.

Finally suppose the operation involves joining two consecutive edges e and e' where $l(e) = r(e')$. Then no critical point is to be specified for the resulting edge e'' because $l(e'') = r(e'')$. Therefore there are several choices for $w(e)$ and $w(e')$. Again up to equivalence, we may assume that $w(e) = w(e')$. We claim that up to equivalence, there are only finitely many choices for $w = w(e) = w(e') \in Z$. In fact, we claim that two choices w_1 and w_2 are equivalent if $pr(w_1) = pr(w_2) \in X$. For, if that is the case, then $w_2 = \gamma w_1$. Then $W_2 = (1, \ldots, \gamma, \gamma, \ldots, 1)W_1$. The term $\nu_J(1 \ldots, \gamma, \gamma, 1)$ is equal to zero, because

$$g_{l(e)r(e)}(\gamma z) + g_{l(e')r(e')}(\gamma z) - g_{l(e)r(e)}(z) - g_{l(e')r(e')}(z) = 0$$

since $l(e) = r(e')$ and $r(e) = l(e')$, so the sum of the first two terms is zero and the sum of the second two terms is zero. This completes the proof that up to equivalence there are only finitely many pairs (W, ϖ) with a given invariant (U, μ).

To complete the proof of compactness, we must show that if W and W' are rosters, and α and α' are maps from J or J' to I respectively, and if the subsidiary given by W and α intersects the subsidiary given by W' and α', then the invariants of $(W, 0)$ and $(W', 0)$ are equivalent.

Let (U, μ) and (U', μ') denote the invariants of W and W' respectively. Suppose that the images of the subsidiaries corresponding to W and W' intersect in Z_I/Γ_I. Then we may translate by Γ_I so that the images intersect in Z_I. This translation can be accomplished by acting on W with Γ_I. This preserves the invariant, so we may assume that the images intersect in Z_I.

Let Y denote the diagram (row of edges and vertices) of length $n = |I|$, labeled with indices corresponding to the index I. The map α corresponds to an order preserving map α^+ from the edges of Y to the edges of W. The points in the subsidiary corresponding to W are given by choices of points $z(e) \in Z$ for the edges of W, such that if $l(e) \neq r(e)$, then $z(e) = w(e)$. The image of such a point is the point $(z_1, \ldots, z_n) \in Z_I$, where $z_k = z(\alpha^+(e_k))$. Thus the condition that the sections corresponding to W and W' intersect is equivalent to the condition that there exist choices $z(e)$ and $z'(e')$ for edges e of W and e' of W', such that for any edge e_k of Y, $z(\alpha^+(e_k)) = z'(\alpha'^+(e_k))$. Construct a new diagram V as follows. An edge of V is a collection of edges of W and W' minimal for the property that if there is an edge e and an edge e' with e' in the collection, and if there is an edge e_k such that $\alpha^+(e_k) = e$ and $\alpha'^+(e_k) = e'$, then e is in the collection; and vice-versa. The indices are assigned to the vertices in an appropriate way, for example an edge of V can be considered as a sequence of consecutive edges in Y, then the indices attached to the end vertices are those attached to the first and last vertices in the sequence of consecutive edges. Now there is a map a from the edges of W to the edges of V, and similarly there is a map a' from the edges of W' to the edges of V. Furthermore, there is a point $v(e)$ attached to every edge of V, in such a way that if e is an edge of W with $l(e) \neq r(e)$, then $w(e) = v(ae)$, and similarly for W'. This is because of the condition that the sections intersect: since the edges of V corresponded to minimal collections of edges of W and W', the points z_k for edges e_k in the sequence of edges of Y corresponding to an edge of V, are all equal.

Now it is clear that we may choose a sequence of steps in the calculation of the invariant of $(W, 0)$, which reduces this pair to the pair $(V, 0)$. Note that since the critical points corresponding to edges within an edge of V are actually equal, there is no need to act by a group element before contracting an edge, so the number ϖ remains equal to zero. Therefore the invariant associated to $(W, 0)$ is the same as the invariant associated to $(V, 0)$. The same is true for W', hence the invariants associated to $(W, 0)$ and $(W', 0)$ are equal.

This completes the proof that the components of the critical point set are compact.

Uniformity of N

To complete the proof of condition (2.5.4), we need to obtain a uniform bound for the numbers N in the expansions of $f_n(\zeta)$. Let U^I be a neighborhood of some component of the critical point set in Z_I/Γ_I. Here everything looks like in the algebraic situation, because the critical point set is a compact subvariety. Let s be the point to which the component maps under g. For any ζ close to s, let $U^I_\zeta = g^{-1}(\zeta) \cap U^I$. For any J and $\alpha : J \to I$ let $U^{J,\alpha} \subset Z_J/\Gamma_J$ denote $\alpha^{-1}U^I$, and similarly for $U^{J,\alpha}_\zeta$. Then there is an action of the monodromy operator T on the homology $H_*(U^I_\zeta, \bigcup_\alpha \alpha U^{J,\alpha}_\zeta)$. There is a number N_I such that $(T^{N_I} - I)^K = 0$ on this homology group for some K. We have to show that there is a uniform bound for this number N_I, independent of the index I or the component of the critical point set.

There is a spectral sequence for relative homology, which converges to $H_*(U^I_\zeta, \bigcup_\alpha \alpha U^{J,\alpha}_\zeta)$. The E^2 term is a direct sum of homology groups of the form $H_*(U^J_\zeta)$. On each of these groups we have $(T^{M_J} - I)^K = 0$ for some K. The number N_I is the least common multiple of the M_J which occur for the terms in the spectral sequence. Thus it suffices to show that there is a bound for the M_J.

Now recall from above that we can write $Z_J = Z^a \times Z^b$ where g is constant on Z^b and has isolated critical points on Z^a. Then $Z_J/\Gamma_J = Z^a/\Gamma' \times X^b$. We may enlarge $U^{J,\alpha}$ until it has the form $U = U' \times X^b$, for a relatively compact $U' \subset Z^a/\Gamma'$. Then $U_\zeta = U'_\zeta \times X^b$. Further we may choose a realization of the monodromy operator which is constant in the X^b direction. By the Künneth formula it suffices to bound the exponent M such that $(T^M - I) = 0$ on $H_*(U'_\zeta)$. In other words, we may assume that g has isolated singularities on U.

Suppose u is a class in $H_*(U_\zeta)$. There is a retraction R from U to U_s, and we may assume that this retraction commutes with the monodromy operator. In particular, $R(T-I)u = 0$. Note that $(T-I)$ is a factor in $(T^M - I)$. Therefore it suffices to bound the number M such that $(T^M - I)^K u = 0$ for some K for all u such that $Ru = 0$. But the singular fiber U_s has isolated singularities, so if $Ru = 0$ then u is a sum of classes supported on small neighborhoods of the singularities.

A singularity has the form $(s_1, \ldots, s_n)$, where s_k are singularities of $g^k = g_{j_{k-1}j_k}$ on Z. Let U^k be a small neighborhood of s_k, and we may assume $U = U^1 \times \ldots U^n$. Choose local coordinates z_k so that s_k is given by $z_k = 0$, and $g^k = (z_k)^{\nu_k}$ (we may change g^k by constants–this moves the point s to 0). The ν_k come from a finite set of numbers, depending on the orders of zeros of the one-forms a_i. Note that the g^k are not identically zero because of the condition that the singularities are isolated.

In terms of the coordinates $(z_1, \ldots, z_k)$ the function g is given by

$$g = z_1^{\nu_1} + \ldots + z_n^{\nu_n}.$$

We have to study the monodromy transformation T on the homology of the intersection of a fiber $g^{-1}(\eta)$ with a small neighborhood of the origin. Let M be the least common multiple of the ν_k. We claim that $(T^M - I)^2 = 0$. For each k set $\mu_k = M/\nu_k$. Introduce new variables t_k and set $z_k = t_k^{\mu_k}$. Then this determines a finite covering of the above space with the function g given by

$$g = t_1^M + \ldots + t_n^M.$$

It suffices to prove that $(T^M - I)^2 = 0$ on the homology of the covering, because we can pull back homology classes to finite covers. Now blow up the origin. The new fiber $g^{-1}(0)$ consists of the strict transform of the old fiber, which is now smooth, and a $\mathbf{P}^{n-1}$ with multiplicity M, meeting the strict transform in a smooth Fermat hypersurface. Proceed with the next step of semistable reduction, which is to pull back by the covering $g = h^M$ of the disc, and then to normalize. Now the fiber $g^{-1}(0)$ is a union of the strict transform of the original fiber, and a smooth variety which is a covering of $\mathbf{P}^{n-1}$ ramified along the Fermat hypersurface. The two components intersect in a smooth Fermat variety. Now we have to show that $(T - I)^2 = 0$. But if u is a homology class, then $(T - I)u$ is a vanishing cycle. It is supported near the intersection of the two components. The fiber near this intersection is topologically a cylinder bundle over the Fermat hypersurface, and the monodromy transformation acts trivially. Therefore $(T - I)^2 u = 0$. This completes the proof of condition (2.5.4).

Condition (2.5.5)

Recall that the statement of Theorem 1 says that given a matrix B of one-forms, we can find a number χ and change B to the matrix χB such that there is a non-zero asymptotic expansion. We therefore must show that we can choose χ so that condition (2.5.5) holds. Notice first of all that the function f_n is homogeneous of degree n in the matrix B, in other words if we change B to χB then f_n changes to $\chi^n f_n$. In particular all of the expansions and continuations still hold. If $c_{jk}^n(s)$ were the coefficients of the expansion of f_n at a point s with respect to the original matrix B, then the new coefficients are $\chi^n c_{jk}^n(s)$. Now note that there are countably many singular points, and hence countably many coefficients altogether. Let $K \subset \mathbf{C}$ be the subfield generated by all of these coefficients, for all singular points. By comparing cardinalities, we may choose a number χ which is transcendental over K. Now let $H_n = \chi^n K \subset \mathbf{C}$. These are independent subspaces, for a dependence among them would give an equation with coefficients in K satisfied by χ. The old coefficients $c_{jk}^n(s)$ are all in K, so the new coefficients $\chi^n c_{jk}^n(s)$ are in H_n. We remark also that the set of possible choices of χ is the complement of a countable set, namely the algebraic closure of K. This completes the proof of (2.5.5).

11. Complements and Examples

In this section we will review what has been proved, make some further comments, and give some examples.

Review

The transport matrix $m(Q,t)$ is given by the infinite sum of integrals

$$m(Q,t) = \sum_n m_n(Q,t), \qquad m_n(Q,t) = \sum_{|I|=n} \int_{\beta_I} be^{tg}.$$

The Laplace transform is again an infinite sum

$$f(\zeta) = \sum_n f_n(\zeta), \qquad f_n(\zeta) = \sum_{|I|=n} \int_{\beta_I} \frac{b}{g-\zeta}.$$

We have shown in §§4-10 that this infinite sum satisfies the conditions (2.5.0)-(2.5.4), and furthermore that if B was multiplied by a generic number χ then it satisfies (2.5.5). Proposition 2.5 shows that $f(\zeta)$ has an extension with locally finite branching and quasi-regular singularities. As noted in the remark following 2.5, the sum of the expansions $\hat{f}_{n,sing}$ converges formally to the expansion $\hat{f}_{sing}$ at any singularity. By 2.3, the transport matrix $m(Q,t)$ has an asymptotic expansion, which is the formal sum of the expansions for $m_n(Q,t)$. If B is multiplied by a generic number, then 2.5 shows that $f(\zeta)$ has faithful expansions, so the asymptotic expansion for $m(Q,t)$ is nonzero. This is the proof of Theorem 1 (and Variant 1.1).

Other cases

The argument described above made use of the hypothesis that S was a compact Riemann surface in three ways:

(*a*) First, and most important, is that the singular metric $\inf_{i\neq j} |dg_{ij}|$ on Z is complete. We used this in §6 to show that the process of moving cycles a finite distance does not push the points off to infinity.

(*b*) Second, the compactness of S means that there is a uniform bound for the multiplicities of the critical points of g_{ij}.

(*c*) Third, we have used the compactness of S in the proof of condition (2.5.4), that the terms in the Picard expansion have regular singularities, and that the number N is uniform.

The method outlined above should be applicable to certain other circumstances, where S may not be compact but these three facts still work. Mainly, suppose that the Riemann surface S is the complex plane, so its universal cover Z is also the complex plane with coordinate z. Suppose that the matrices of one-forms A and B have entries which are polynomials in z, times dz. The integral of a polynomial is again a polynomial, so $g_i(z)$ are polynomial functions on Z. Condition (a) still works because any nonzero polynomial one-form on the complex plane has a pole at infinity. Thus the metric $\inf_{i\neq j} |dg_{ij}|$ is still complete. The second condition (b) is clear since there are only finitely many critical points of the g_{ij}. The third condition (c) works because the functions $g_I : Z_I \rightarrow \mathbf{C}$ are polynomials. In this algebro-geometric situation, integrals such as $f_I(\zeta) = \int_{\beta_I} \frac{b}{g-\zeta}$ have regular singularities (due to the regular singularity of the Gauss-Manin connection). Admittedly, one would have to do some work to obtain a uniform bound for N. If this should fail, we would still obtain asymptotic expansions, with the denominator N increasing with the power of t^{-1}.

What about extending the method to the case of other quasi-projective Riemann surfaces? Condition (b) is no problem, and condition (c) should be possible to treat with some additional work. The main restriction is that condition (a) will work if and only if the one-forms dg_{ij} have nontrivial poles at all of the points at infinity, for all $i \neq j$. This may or may not be the case in any given example, and if it is not so then the situation becomes more complicated. We can formulate this as a more difficult problem for future study:

What happens when the matrix B has poles at points where some $a_i - a_j$ are regular?

The first part of our method should still work in this case—it should be possible to choose flows which avoid the singularities of B, and show that the Laplace transform of the solution has an extension with locally finite branching. But it seems that the Laplace transform might have essential singularities. Then it would not be so clear what type of asymptotic information to look for.

Convolutions

The operation of Laplace transform takes multiplication to convolution. Suppose $f(\zeta)$ and $g(\zeta)$ are two functions defined for $|\zeta| \geq a$. Their *convolution* is the function

$$f * g(\zeta) = \frac{1}{2\pi i} \oint f(x) g(\zeta - x) dx.$$

For large values $|\zeta| > 2a$, the singularities of $f(x)$ and $g(x-\zeta)$ are contained in the disjoint sets $|x| \leq a$ and $|\zeta - x| \leq a$ respectively. The contour of integration is then taken around one of these groups of singularities, for example around $|x| = a + \epsilon$.

Lemma 11.1 *If $m(t)$ and $n(t)$ are entire functions of order ≤ 1, with Laplace transforms $f(\zeta)$ and $g(\zeta)$, then the Laplace transform of $m(t)n(t)$ is $f * g(\zeta)$. If f and g have analytic continuations with locally finite regular singularities, then $f * g$ has an analytic continuation with locally finite regular singularities.*

Proof:

$$\frac{1}{2\pi i}\oint f * g(\zeta)e^{\zeta t}d\zeta = \frac{1}{(2\pi i)^2}\int\int f(x)e^{xt}g(\zeta - x)e^{(\zeta - x)t}dx d\zeta$$

where the double integral is taken over the cycle $|x| = a + \epsilon$, $|\zeta| = 4a$. We can interchange the order of integration and integrate $d\zeta$ first. For a given value of x with $|x| = a + \epsilon$,

$$\int_{|\zeta|=4a} g(\zeta - x)e^{(\zeta - x)t}d\zeta = \int_{|\zeta - x|=4a} g(\zeta - x)e^{(\zeta - x)t}d(\zeta - x) = 2\pi i\, n(t).$$

Then

$$\frac{1}{(2\pi i)^2}\int_{|x|=a+\epsilon} 2\pi i n(t) f(x)e^{xt}dx = m(t)n(t).$$

The function $f * g(\zeta)$ vanishes at infinity, so it is the Laplace transform of $m(t)n(t)$.

Suppose $f(\zeta)$ and $g(\zeta)$ have locally finite branching, with sets of singularities denoted S'_M and S''_M respectively. If ζ is moved along a path in $\mathbf{C}$ then the path of integration in the expression

$$f * g(\zeta) = \frac{1}{2\pi i}\oint f(x)g(\zeta - x)dx$$

may be deformed continuously so as to miss the singularities of $f(x)$ and $g(\zeta - x)$. If ζ is moved along a path of length $\leq M$, then no point of the path of integration needs to be moved a distance greater than M, and similarly the value of $\zeta - x$ at a point x on the path of integration will not be moved a distance greater than M. So in this process, the only singularities which come into play are those at $x \in S'_M$ and $\zeta - x \in S''_M$. As long as these two cases do not occur at once, the path of integration may be kept away from the singularities.

So the moving process works as long as ζ is moved along a path of length less than or equal to M which stays away from

$$S_M = \{s = s' + s'' : s' \in S'_M,\ s'' \in S''_M\}.$$

So $f * g(\zeta)$ may be analytically continued along any such path, showing that it has locally finite branching.

Now use the hypothesis that f and g have regular singularities. Fix a point $s \in S_M$. There are finitely many pairs $(s', s'') \in S'_M \times S''_M$ with $s' + s'' = s$. For ζ near s, the path of integration may be deformed to a union of segments of the following kinds. There may be segments which stay away from $x = s'$ for $s' + s'' = s$. These segments give a holomorphic contribution to $f * g(\zeta)$ at s, so we can ignore them. The segments which contribute have the form

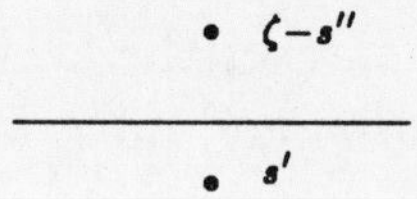

γ.

As $\zeta \to s$, the two singularities $x = s'$ and $x = \zeta - s''$ move together. We will show that the contribution

$$h(\zeta) = \int_\gamma f(x) g(\zeta - x) dx$$

has a regular singularity at $\zeta = s$. When moving ζ, keep the endpoints of the path γ fixed. It is clear that $h_1(\zeta)$ has polynomial growth, since both terms in the integrand have polynomial growth near their singularities. Recall that T denotes the operation of continuing a function around the singularity of its argument. Apply the criterion of Lemma 2.2. From the picture,

$$(T - I) h(\zeta) = \int_\alpha f(x) g(\zeta - x) dx$$

where α is the closed path

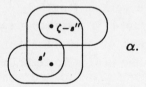

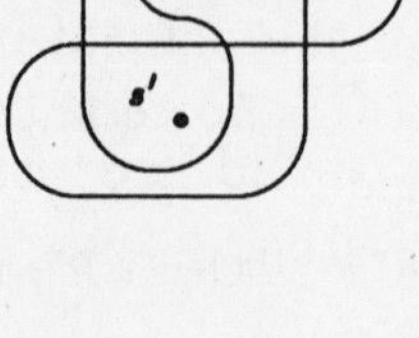

α.

Now when ζ is moved around s, the closed path α can be dragged along, so the only effect is to continue the integrands around their singularities. Thus

$$T^k(T-I)h(\zeta) = \int_\alpha (T^k f)(x)(T^{-k}g)(\zeta - x)dx.$$

Formally, the tensor product of two quasi-unipotent transformations is again quasi-unipotent. So if $(T_1^{N_1} - I)^{K_1} f = 0$ and $(T_2^{N_2} - I)^{K_2} g = 0$ then there are N and K such that $((T_1 \otimes T_2)^N - I)^K f \otimes g = 0$. The numbers N and K depend universally on N_1, N_2, K_1, and K_2, independently of the spaces on which T_1 and T_2 act. The latter identity is a formal consequence of the first two. We can apply the proof of this identity to the case where T_1 represents the action of T on f and T_2 represents the action of T^{-1} on g. Both of these operations are quasi-unipotent by hypothesis, so there are N and K with

$$(T^N - I)^K (T-I)h(\zeta) = 0.$$

Thus h has a regular singularity at $\zeta = s$. Adding these contributions for the various segments passing near points s', we conclude that $f * g(\zeta)$ has a regular singularity at $\zeta = s$.

Exercise: Show that if f and g have quasi-regular singularities, then $f * g$ has quasi-regular singularities.

Let $\mathbf{Q}(\Gamma)$ denote the field generated over $\mathbf{Q}$ by the roots of unity, π, and the values and derivatives of the Gamma function at rational points. The new coefficients arising from convolutions of powers of ζ and $\log \zeta$ lie in $\mathbf{Q}(\Gamma)$. This may be seen as in the proof of 2.3.

Proposition 11.2 *Suppose that* $f = \sum f_n$ *and* $g = \sum g_n$ *satisfy conditions* (2.5.0) *through* (2.5.4) *from §2. Then the convolution* $f * g = \sum f_m * g_n$ *satisfies those conditions. If* f *and* f *satisfy condition* (2.5.5) *with respect to vector spaces* H'_n *and* H''_n, *and if there are vector spaces* H_k, *independent over* $\mathbf{Q}(\Gamma)$, *such that* $H'_m H''_n \subset H_{n+m}$, *then* $f * g$ *satisfies condition* (2.5.5).

Proof: Conditions (2.5.0), (2.5.1), and (2.5.2) are clear. Condition (2.5.4) follows from the previous lemma (the uniformity of the exponent N follows from the proof of 11.1 above).

For condition (2.5.3), note that we can safely ignore the factor of $1/k!$, absorbing it in $n^{-\epsilon n}$ at the expense of making ϵ smaller (the condition was stated with the factorial in order to make it more convenient to put directly into Taylor's formula). We can extend the bounds (2.5.3) to negative values of k, by choosing a point ζ_0 and setting

$$f^{(-1)}(\zeta) = \int_{\zeta_0}^{\zeta} f(x)dx$$

and iterating to get $f^{(k)}$ for $k < 0$. Then (2.5.3) implies that for all $-b \leq k \leq \varepsilon(k)n - b$,

$$|f_n^{(k)}(\zeta)| \leq C(k)^n n^{-\varepsilon(k)n}$$

uniformly for ζ in a sector near a singularity s', with constants $C(k)$ and $\varepsilon(k)$ depending on k but not n. We get a similar bound for $g(\zeta)$ near a singularity s'' (choose constants b, $C(k)$, and $\varepsilon(k)$ which work for both f and g).

Fix $k \geq 0$, and look at a contribution

$$h_{m,n}^{(k)} = \frac{d^k}{d\zeta^k} \int_\gamma f(x)g(\zeta - x)dx.$$

where γ is a segment passing between $x = s'$ and $x = \zeta - s''$ as in 11.1. Let ε be the smallest of $\varepsilon(k')$ for $-b \leq k' \leq k + b$, and let C be the largest of the $C(k')$. Suppose m and n are integers with $k \leq \varepsilon(m + n) - 2b$. Then there are k' and k'' with $k' + k'' = k$, $k', k'' \geq -b$, and $k' \leq \varepsilon m - b$ and $k'' \leq \varepsilon n - b$.

Integrating by parts,

$$h_{m,n}^{(k)}(\zeta) = H_{m,n}^{k',k''}(\zeta) + \int_\gamma f_m^{(k')}(x) g_n^{(k'')}(\zeta - x).$$

Here $H_{m,n}^{k',k''}(\zeta)$ is a sum of products of derivatives of $f_m(\zeta)$ and $g_n(\zeta)$ evaluated at the endpoints of γ. This may be bounded by $C^{m+n}(m+n)^{-\varepsilon(m+n)}$ using condition (2.5.2). On the other hand the integrand is bounded by $C^{n+m} n^{-\varepsilon n} m^{-\varepsilon m}$, using the conditions (2.5.3) (possibly as extended above to negative k). By increasing C and decreasing ε, this will be bounded by $C^{n+m}(m + n)^{-\varepsilon(m+n)}$, which gives the bound (2.5.3).

Finally, under the additional hypotheses in the last part of the lemma, we get condition (2.5.5), because the coefficients in the singular parts of the expansions for $f * g$ are products of the coefficients for f and g, with some additional coefficients which are in $\mathbf{Q}(\Gamma)$. The additional coefficients come from integrals of the form

$$\frac{\partial^{k+l}}{\partial a^k \partial b^l} \int x^a (\zeta - x)^b dx.$$

Such an integral is equal to a derivative of ζ^{a+b+1} times the Beta function.

Let us return to the family of differential equations

$$(d - tA - \chi B)m = 0.$$

Here χ is a finite generic parameter.

Proposition 11.3 *Suppose $\Phi(t)$ is a polynomial in derivatives of the matrix coefficients $m_{ij}(Q,t)$, with coefficients which are polynomials in t. Let $f(\zeta)$ be the Laplace transform of $\Phi(t)$. Then $f(\zeta)$ has an analytic continuation with locally finite branching and quasi-regular singularities. If χ is chosen generically, then $f(\zeta)$ has faithful expansions. Consequently $\Phi(t)$ has an asymptotic expansion as $t \to \infty$ in any given direction. The asymptotic expansion is nonzero if χ is generic and $\Phi(t)$ is not identically zero.*

Proof: The Laplace transform of a polynomial in t is a meromorphic function with some poles at the origin only. Similarly, taking the derivative of a function corresponds to multiplying its Laplace transform by ζ (and subtracting off the appropriate constant to maintain vanishing at infinity). This preserves the conditions 2.5. By Lemma 11.1, the Laplace transform $f(\zeta)$ of the polynomial $\Phi(t)$ will be a sum of convolutions of Laplace transforms of polynomials in t, and Laplace transforms of matrix coefficients $m_{ij}(t)$ (or their derivatives). Let $\mathbf{k}$ denote the subfield generated over $\mathbf{Q}(\Gamma)$ by all coefficients in the case $\chi = 1$. Then assume that χ is transcendentally independent of $\mathbf{k}$, and let $H_n = \chi^n \mathbf{k} \subset \mathbf{C}$. The vector spaces H_n are independent over $\mathbf{Q}(\Gamma)$, $H_m H_n \subset H_{n+m}$, and the coefficients of the expansions satisfy condition (2.5.5) with respect to these H_n (because if $f(\zeta)$ is a Laplace transform of a matrix coefficient, then the integrands of the terms $f_n(\zeta)$ are homogeneous of degree n in the matrix B). Apply Propositions 2.5 and 11.2 to obtain the conclusions.

Remark: The same statement and proof hold if the coefficients of Φ are functions whose Laplace transforms have locally finite regular singularities.

Remark: One would like to show a strong transcendence statement, namely that the matrix coefficients $m_{ij}(Q,t)$ do not satisfy any differential equation, even with exponential functions (or functions whose Laplace transforms have finitely many regular singularities) as coefficients. The analytic continuations of the Laplace transforms that we have obtained should be helpful here. The basic problem remains to figure out a good method of calculating the locations of the singularities and the coefficients of the quasi-regular expansions.

If Q is a point on the universal cover Z which differs from P by following a closed loop $\gamma \in \pi_1(S,P)$, then the solution $m(\gamma,t) = m(Q,t)$ is the monodromy matrix for γ. The family of differential equations corresponds to an algebraic curve in the moduli space of vector bundles with integrable connections, the *de Rham moduli space* $\mathbf{M}_{DR}$ of [29]. The map taking a connection to its monodromy representation is a map from $\mathbf{M}_{DR}$ to $\mathbf{M}_B$, the *Betti moduli space* of representations of the fundamental group considered as an abstract discrete group. The Betti moduli space is an affine variety. By the theorem of Weyl-Procesi [28], the polynomial functions on $\mathbf{M}_B$ are sums of monomials of the form

$$Tr(m(\gamma_1))Tr(m(\gamma_2))\cdots Tr(m(\gamma_n))$$

where $m(\gamma_i)$ denote the matrices representing various loops in the fundamental group. We can interpret our asymptotic expansions as giving information on the asymptotic behaviour of any polynomial function on $\mathbf{M}_B$, pulled back to a holomorphic function of t on the curve of connections.

Pathwise bound for the exponent

There is an easy bound for the exponent ξ of the asymptotic expansion, obtained by looking at a path from P to Q. Suppose $\gamma : [0,1] \to Z$ with $\gamma(0) = P$ and $\gamma(1) = Q$. Set

$$\xi_{path}(\gamma) = \int_0^1 \sup_i (\Re\gamma^* a_i)(t).$$

Note that the pullbacks $\gamma^* a_i$ are one-forms $\alpha_i(t)dt$. For each t, the integrand $\sup_i(\Re\gamma^* a_i)(t)$ is the one-form $(\sup_i \Re\alpha_i(t))dt$. For different values of t along the path, the supremum will be attained with different values of i. Set

$$\xi_{path} = \inf_\gamma \xi_{path}(\gamma).$$

This infimum is attained for some path γ.

Lemma 11.4 *Suppose ξ is the real exponent of the asymptotic expansion for $m(Q,t)$. Then $\xi \leq \xi_{path}$.*

Proof: Form the cycle of integration β_* corresponding to the path γ which attains the infimum $\xi_{path}(\gamma) = \xi_{path}$. If $z = (z_1,\ldots,z_n) \in Z_I$ is a point in the support of β_*, it means $z_i = \gamma(t_i)$ for $0 \leq t_1 \leq \ldots \leq t_n \leq 1$. Then

$$\Re g(z) = \int_0^{t_1} \Re\gamma^* a_{i_0} + \ldots + \int_{t_n}^1 \Re\gamma^* a_{i_n}$$

is bounded by $\xi_{path}(\gamma) = \xi_{path}$. Therefore $Supp_{\mathbf{C}}(\beta_*)$ is contained in the region $\Re x \leq \xi_{path}$. Hence the Laplace transform

$$f(\zeta) = \int_{\beta_*} \frac{b}{g - \zeta}$$

is analytic in $\Re\zeta > \xi_{path}$. This implies that the exponent of the asymptotic expansion for $m(t)$ is less than ξ_{path} (see Proposition 2.3).

Remark: In fact, $\xi_{path}(\gamma)$ is the best possible bound which can be obtained from the information of the system of differential equations along γ (at least without making some analytic continuation, in effect looking away from γ).

We will see that in the case of $S\ell(2)$ connections, the exponent ξ is equal to ξ_{path}. However, for $S\ell(n)$ connections with $n \geq 3$, ξ will usually be smaller than ξ_{path}. This will be demonstrated by an example. This example shows that our asymptotic expansion cannot be obtained simply by looking at the behaviour near one path γ.

Here is a proposition which will be useful for showing that $\xi < \xi_{path}$.

Proposition 11.5 *Suppose γ is a path from P to Q such that $\xi_{path}(\gamma) = \xi_{path}$ is attained. Suppose that γ does not pass through any critical points of g_{ij}, and suppose that $\xi_{path}(\gamma) > \Re g_i(Q)$ for all i. Then $\xi < \xi_{path}$.*

Proof: Use the path γ to define an initial cycle of integration β_*. As in the previous lemma, $Supp_{\mathbf{C}}(\beta_*)$ is contained in the half-plane $\Re x \leq \xi_{path}$. The fact that $\xi_{path}(\gamma)$ is minimized means that γ does not intersect itself. Let U denote a contractible neighborhood of γ which contains no critical points of g_{ij}. Choose a flow f^0 so that it retracts U to a point inside U. The first claim is that there is a number $\mu > 0$ such that if $z = (z_1, \ldots, z_n) \in Z_I$ is a point beyond points of β_* at distance $\leq \mu$, then $z_k \in U$ for $k = 1, \ldots, n$. This can be proved as in §6, with a function ϕ_1, defined on U, replacing the function ϕ_0 in the first few paragraphs of the proof of Proposition 6.1. The condition that the flow f^0 retracts U inside itself means that we can insure that there is a number μ with $\sup_\gamma \phi_1 + \mu < \inf_{\partial U} \phi_1$. Then if z is beyond points of β_* at distance less than μ, we have $\phi_1(z_k) \leq \sup_\gamma \phi_1 + \mu$, so $z_k \in U$. In particular, if $\ell = (\ell_1, \ldots, \ell_n)$ is a critical point beyond β_* at distance $\leq \mu$, then $n = 0$. In this trivial case, $Z_I = Z_{(i_0)} = \{\ell\}$ is a point, automatically considered as a critical point, with $g_I(\ell) \stackrel{def}{=} g_{i_0}(Q)$. Under the hypotheses of this proposition, we can make μ smaller so that $\Re g_i(Q) \leq \xi_{path} - \mu$ for all i. Then $S_\mu = \{g_i(Q)\}$ is contained in $\Re x \leq \xi_{path} - \mu$.

Now choose flows f_{ij} with respect to $L = \mu$ and small choices of δ and σ. Apply the moving procedure outlined in §8, but with η_1 replaced by the full chain of integration β_*. There are no boundary terms, and no critical points are encountered in the region $\Re x \geq \xi_{path} - \mu$. After making the appropriate cutoffs, β_* becomes homologous to a pro-chain η' satisfying the bounds $\mathbf{F}$, and supported in $\Re g(z) \leq \xi_{path} - \mu$. This provides an analytic continuation of $f(\zeta)$ to $\Re\zeta > \xi_{path} - \mu$, showing that the exponent ξ of the asymptotic expansion is strictly smaller than ξ_{path}.

$S\ell(2)$ connections

In the case of connections with structure group $S\ell(2)$, we will give a topological interpretation of the real part of the exponent $\xi = \Re\lambda$ in the asymptotic expansion (essentially, $\xi = \xi_{path}$). This interpretation shows that the exponent has some interesting geometric significance, and it suggests that it would be interesting to explore the significance in the case of connections of higher rank.

The condition that the structure group is $S\ell(2)$ means that the bundle in question has rank 2. Thus the matrices A and B are 2×2. The condition that the connection has determinant one translates infinitesimally to the condition that the matrix of one-forms has trace zero, in other words $Tr(A) = 0$. Note that $Tr(B) = 0$ automatically since $b_{ii} = 0$. Thus our matrices of one-forms contain three one-forms, a, b_{12}, and b_{21}:

$$A = \begin{pmatrix} a & 0 \\ 0 & -a \end{pmatrix}$$

$$B = \begin{pmatrix} 0 & b_{12} \\ b_{21} & 0 \end{pmatrix}.$$

We will eventually need to assume that B is sufficiently generic with respect to A. To the one-form a there corresponds a function $g(z) = \int_P^z a$ on the universal cover Z of the Riemann surface. Thus in our previous notation, $a_1 = a$ and $a_2 = -a$, so $g_1 = g$ and $g_2 = -g$.

The simplification in the case of $S\ell(2)$ connections comes from the fact that we can choose our path γ so that the resulting iterated integrals already have stationary phase form, and we do not need to move the chains of integration.

To be more precise, define a singular Finsler metric on S or Z by

$$ds = |d\Re g|.$$

In particular if $\gamma : [0,1]$ is a path then its length in this metric is

$$\xi_{path}(\gamma) = \int_0^1 |d\Re g(\gamma(t))/dt|dt.$$

This measures the total variation in $\Re g$ along the path γ.

The geodesics in this metric are not unique, but they do exist. If P and Q are two points in Z, and if a is generic with respect to P and Q, then there is a path γ from P to Q which minimizes length in the metric $|d\Re g|$, and which has the property that any local maxima or minima of $\Re g$ along the path are attained at critical points of $\Re g$ on Z. Furthermore we may assume that near such points, γ is a path of steepest descent or ascent.

Proposition 11.6 *Let $\xi = \xi_{path}(\gamma)$ be the length of the geodesic γ from P to Q. Suppose a and B are generic. Then in the asymptotic expansion for the monodromy $m(t)$ for positive real $t \to \infty$, the real parts of the exponents λ_i are equal to ξ.*

Proof: By Lemma 11.4, the real part of the exponent for each $m_n(t)$ is $\leq \xi$. We will show that the real part of the exponent is equal to ξ for some $m_n(t)$. This will prove the proposition since B is generic (see Variant 1.1). Fix a geodesic path γ from P to Q. If $0 \leq t_1 \leq \ldots \leq t_n \leq 1$ and $z = (\gamma(t_1), \ldots, \gamma(t_n))$ then for any $I = (i_0, \ldots, i_n)$ we have

$$g_I(z) = \sum_{j=0}^{n} (g_{i_j}(\gamma(t_{j+1})) - g_{i_j}(\gamma(t_j))),$$

with the convention that $t_0 = 0$ and $t_{n+1} = 1$. Each g_i is either g or $-g$. Because of the condition that a is generic, the relative maxima and minima of the function $\Re g$ along the geodesic are isolated, and they alternate. Let n be the number of them, and let $s_1, \ldots, s_n$ be the relative maxima and minima. Suppose that s_1 is a maximum (resp. minimum). Choose I with $|I| = n$ to be $I = (1,2,1,2,\ldots)$ (resp. $I = (2,1,2,1,\ldots)$). Thus

$$\frac{d}{dt} \Re g_{i_j}(\gamma(t)) \geq 0$$

for $t \in [t_j, t_{j+1}]$. Then $\Re g_I(\gamma(s_1), \ldots, \gamma(s_n)) = \xi$.

We claim that for any index J with $|J| \leq n$ but $J \neq I$, $\sup_{\beta_I} \Re g_J(z) < \xi$. Note first of all that if there are repeated 1's or 2's, then we may replace J

by an index of smaller length where there are no repititions, but which gives the same supremum. Thus either $J = (1,2,1,2,\ldots)$ or $J = (2,1,2,1,\ldots)$. Let $(t_1,\ldots,t_m)$ be the point where the supremum is achieved. If $t_k = t_{k+1}$ then we may remove the index j_k and the point t_{k+1} without affecting the value of the supremum; and then remove repetitions again as before. Thus we may assume the t_i are distinct. Similarly, we may assume that they are distinct from 0 or 1. Next, we may assume that t_k is a relative maximum for $\Re(g_{j_{k-1}} - g_{j_k})$ (otherwise we increase the value of $\Re g_J$ by moving t_k). In particular, t_k is either a relative maximum or relative minimum of h, so $t_k = s_l$ for some l. Furthermore we chose I so that s_l is a relative maximum for $\Re(g_{i_{l-1}} - g_{i_l})$, so $j_{k-1} = i_{l-1}$ and $j_k = i_l$. Since $J \neq I$, we must now have $|J| < n$. Thus there is a segment $[t_k, t_{k+1}]$ which contains more than one segment $[s_l, s_{l+1}]$. In particular,

$$\Re g_{j_k}(\gamma(t_{k+1})) - \Re g_{j_k}(\gamma(t_k)) < \int_{t_k}^{t_{k+1}} |\frac{d}{dt}\Re g(\gamma(t))|dt.$$

Therefore $\Re g_J(\gamma(t_1),\ldots,\gamma(t_m)) < \xi$. This proves the claim.

In light of the claim, it suffices to prove that $m_I(t)$ has a nonzero asymptotic expansion with real part of the exponent equal to ξ. Let U_i be disjoint open neighborhoods of the points s_i, $i = 1,\ldots,n$. Our next claim is that there is $\xi_1 < \xi$ where if $(t_1,\ldots,t_n)$ is a point such that for some k, t_k is not in U_k, then $\Re g_I(\gamma(t_1),\ldots,\gamma(t_n)) \leq \xi_1$. To prove this, first note that the complement of $U_1 \times \ldots \times U_n$ in $[0,1]^n$ is closed, so $\Re g_I$ attains its maximum on the complement of $U_1 \times \ldots \times U_n$. Thus it suffices to show that if $(t_1,\ldots,t_n) \neq (s_1,\ldots,s_n)$ then $\Re g_I(\gamma(t_1),\ldots,\gamma(t_n)) < \xi$. But if $(t_1,\ldots,t_n) \neq (s_1,\ldots,s_n)$, then there is an interval $[t_k, t_{k+1}]$ which overlaps more than one consecutive interval $[s_l, s_{l+1}]$, $0 \leq k \leq n$ and $0 \leq l \leq n$, with the conventions $t_0 = s_0 = 0$ and $t_n = s_n = 1$. Thus again

$$\Re g_{i_k}(\gamma(t_{k+1})) - \Re g_{i_k}(\gamma(t_k)) < \int_{t_k}^{t_{k+1}} |\frac{d}{dt}\Re g(\gamma(t))|dt,$$

and hence $\Re g_I(\gamma(t_1),\ldots,\gamma(t_n)) < \xi$. This proves the claim.

A consequence of the claim is that the integral of $b_I e^{tg_I}$ over the complement of $U_1 \times \ldots \times U_n$ in β_n grows like $e^{t\xi_1}$ or slower. Hence it suffices to show that

$$\int_{U_1 \times \ldots \times U_n} b_I e^{tg_I}$$

has a nonzero asymptotic expansion with real part of the exponent equal to ξ. But this integral is equal to

$$e^{tg_{i_n}(Q)} \prod_{r=1}^{n} \int_{U_r} b_{i_r i_{r-1}}(\gamma(t)) e^{tg_{i_{r-1} i_r}(\gamma(t))}$$

where $g_{ij} = g_i - g_j$. Thus it suffices to show that the integral

$$\int_{U_r} b_{i_r i_{r-1}}(\gamma(t)) e^{tg_{i_{r-1} i_r}(\gamma(t))}$$

has an asymptotic expansion where the exponent is $\lambda = g_{i_{r-1} i_r}(\gamma(s_r))$. To do this, we use the fact that $\gamma(s_r)$ is a critical point of $g_{i_{r-1} i_r}$, and that near s_r, γ is a path of steepest descent for $\Re g_{i_{r-1} i_r}$. Since a is generic, we may choose a local holomorphic coordinate z at $\gamma(s_r)$ so that $z(\gamma(s_r)) = 0$ and so that $g_{i_{r-1} i_r} = \lambda - z^2$. The path γ is a segment of the real axis in the coordinate z, say from $-\varepsilon$ to ε. Now write $b_{i_r i_{r-1}} = \sum f_k z^k dz$. Then the integral in question is

$$e^{\lambda t} \int_{-\varepsilon}^{\varepsilon} \sum f_k z^k e^{-tz^2} dz.$$

This has an asymptotic expansion given by

$$e^{\lambda t} \sum C_k f_k t^{-(k+1)/2}$$

where

$$C_k = \int_{-\infty}^{\infty} u^k e^{-u^2} du.$$

This is equal to zero for odd values of k, but nonzero for even values. If $b_{i_r i_{r-1}}$ is chosen generically with respect to a, then the term f_0 does not vanish. Therefore the asymptotic expansion is nonzero. This completes the proof of the theorem.

Remark: It is evident from this proposition that for $S\ell(2)$, the asymptotic expansion can be obtained by continuing along a geodesic path. Thus the $S\ell(2)$ case is fully covered by the known method of matched asymptotic expansions. The points $s_1, \ldots, s_n$ encountered above are the "turning points".

As a corollary of 11.6 in the $S\ell(2)$ case, we obtain a description of the convex hull of the exponents $\mathcal{H}$ (defined in §2). Fix a generic choice of one-form a. Consider the resulting function $g : Z \to \mathbf{C}$ as a branched covering. Given points P and Q in Z, there is a path γ between them with the following properties:

$g_*(\gamma)$ is a union of line segments between points $s_0 = g(P), s_1, \ldots, s_r = g(Q)$; the points s_i are the images of branch points; and the path γ changes branches at each corner between segments. Such a path γ may be obtained by choosing a geodesic with respect to the pull-back of the euclidean metric on $\mathbf{C}$.

Corollary 11.7 *With the above notations let $w_i = s_i - s_{i-1}$ be the vector pointing in the direction of the ith segment. For each choice of signs $\nu = (\nu_1, \ldots, \nu_r)$ with $\nu_i = \pm 1$, consider the point*

$$h_\nu = g(P) + \nu_1 w_1 + \ldots + \nu_r w_r$$

in $\mathbf{C}$. Then if B is generic, the convex hull of exponents $\mathcal{H}$ for $m(Q,t)$ is equal to the convex hull of the points h_ν.

Proof: Apply Proposition 11.6 to the case of one-form $ae^{i\theta}$ and matrix B; for generic values of θ the one-form will still be generic. The solution in this case will be $m(Q, e^{i\theta}t)$ where m denotes the solution for a and B. The path γ we have chosen above will work as a path for applying 11.6, for any value of θ. The expression for the real exponent given in Propostion 11.6 is the same as $\xi = \sup_{h\in\mathcal{H}} \Re e^{i\theta} h$, where $\mathcal{H}$ is the convex set defined above. Therefore $\mathcal{H}$ is the convex hull of exponents of $m(Q,t)$.

Remark: This simple description of the hull of exponents might make it possible to solve the inverse scattering problems described in the introduction, for the case of $S\ell(2)$ connections.

Example: some $S\ell(3)$ connections

The special behaviour shown by the above example lies in marked contrast to the case of matrices of higher rank. When there is more than one function g, it is no longer possible to choose one path which will exhibit the expansion. We will use the criterion of Proposition 11.5 to construct a 3×3 example where $\xi < \xi_{path}$.

Let $A = \begin{pmatrix} a_1 & 0 & 0 \\ 0 & a_2 & 0 \\ 0 & 0 & -a_1 - a_2 \end{pmatrix}$. Let U be a neighborhood in the Riemann surface S, with a local coordinate $z : U \to \mathbf{C}$ mapping U isomorphically to the disc of radius 5 at the origin. By choosing a_1 and a_2 appropriately, and adjusting the local coordinate, we may assume that

$$a_1 = iC_1 dz$$

$$a_2 = 2(z-i)dz + u(z)dz$$

where C_1 is a large positive real constant, and $|u(z)| \leq c_2$ where $c_2 > 0$ is small. Note that a_2 has a single zero in U, located in $|z-i| \leq c_2$.

Let P and Q be the points $z = -1$ and $z = 1$ respectively, and let γ_0 be the straight line segment joining P to Q. Then

$$\xi_{path}(\gamma_0) \leq 2\varepsilon + \int_{-1}^{1} |\Re 2(z-i)|dz = 2c_2 + 2,$$

independently of C_1. Suppose γ is a path from P to Q with $\xi_{path} = \xi_{path}(\gamma)$. In particular, $\xi_{path}(\gamma) \leq \xi_{path}(\gamma_0)$.

Lemma 11.8 *Suppose $\epsilon > 0$. We can choose C_1 large enough such that γ must stay within U and within the strip where the imaginary part is bounded $|\Im z| \leq \epsilon$.*

Proof: If R is any point on the path γ, then

$$\int_P^R a_i \leq \int_P^Q \sup\{\gamma^* a_1, \gamma^* a_2, \gamma^*(-a_1 - a_2)\} = \xi_{path}(\gamma)$$

for $i = 1, 2$. Similarly,

$$\int_P^R (-a_1) = \int_P^R a_2 + \int_P^R (-a_1 - a_2) \leq 2\xi_{path}(\gamma)$$

and the same for a_2. Therefore

$$\left| \int_P^R a_1 \right| \leq 2\xi_{path}(\gamma).$$

This implies that if C_1 is large, then there can be no point $R \in U$ on the path γ with imaginary part of R greater than ϵ or less than $-\epsilon$. Suppose γ contained a point not in U. Then there would be a point R on γ near the edge of U, with $|\Im R| \leq \epsilon$. The radius of U has been chosen large enough so that this contradicts the estimate

$$\left| \int_P^R a_2 \right| \leq 2\xi_{path}(\gamma) \leq 4 + 4c_2.$$

This proves the lemma.

Now if C_1 is chosen large, there are no zeros of $a_1 - a_2$, $a_1 - (-a_1 - a_2)$, or $a_2 - (-a_1 - a_2)$ in the strip $|\Im z| \leq \epsilon$ in U. Therefore there are no critical points of g_{ij} along γ. To complete the hypotheses of 11.5 in our example, we must show that $\xi_{path}(\gamma) > \Re g_i(Q)$ for $i = 1, 2, 3$. There is a point R on the path γ with $\Re R = 0$, and $|\Im R| \leq \epsilon$. Then

$$\xi_{path}(\gamma) \geq \int_P^R a_2 - \int_R^Q a_2 \geq \int_P^R 2(z-i)dz - \int_R^Q 2(z-i)dz - 2c_2.$$

Now

$$\int_P^R 2(z-i)dz = (z-i)^2|_P^R = -(1-\Im R)^2 - (-1-i)^2 = -(1-\Im R)^2 - 2i,$$

and similarly

$$\int_R^Q 2(z-i)dz = (z-i)^2|_R^Q = (1-i)^2 + (1-\Im R)^2 = (1-\Im R)^2 - 2i.$$

Now recall that $|\Im R| \leq \epsilon$, so

$$\xi_{path}(\gamma) \geq 2(1-\epsilon)^2 - 2c_2.$$

On the other hand, $g_1(Q) = 0$ and

$$\Re g_2(Q) = \Re \int_P^Q 2(z-i)dz + \Re \int_P^Q u(z)dz = \Re \int_P^Q u(z)dz,$$

so $|\Re g_2(Q)| \leq 2c_2$. Since $g_3(Q) = -g_1(Q) - g_2(Q)$, we have $|\Re g_3(Q)| \leq 2c_2$. If c_2 is chosen small enough, and if C_1 is big enough so that ϵ is small, then

$$\xi_{path}(\sigma) \geq 2(1-\epsilon)^2 - 2c_2 > 2c_2 \geq |\Re g_i(Q)|.$$

We may now apply the criterion of Proposition 11.5 to conclude that, in this example, the actual exponent ξ in the asymptotic expansion is strictly less than the best pathwise bound ξ_{path}. This shows that the asymptotic expansion cannot be obtained by simply continuing the solution along the path.

12. The Sturm-Liouville problem

We formulate a boundary value problem. Let n denote the rank of the system of differential equations. Let $\mu_1(u,v),\ldots,\mu_n(u,v)$ be linear functionals on the space $\mathbf{C}^n \oplus \mathbf{C}^n$. The points P and Q in the universal cover Z are fixed. These determine a boundary value problem (following the description in [3]): to find a vector valued solution $u(z)$ of the system of differential equations

$$(d - tA - B)u = 0,$$

satisfying the boundary conditions

$$\mu_i(u(P), u(Q)) = 0.$$

For which values of t does there exist a nonzero solution u? In terms of the transport matrix $m(Q,t)$, the question becomes: for which values of t does there exist a nonzero vector $u \in \mathbf{C}^n$ such that

$$\mu_i(u, m(Q,t)u) = 0?$$

There is a matrix $M(m)$ such that $\mu_i(u, mu) = M(m)_{ij}u_j$, and the entries of $M(m)$ are linear in the entries of m. The condition that there is a nonzero solution u is just that $\det M(m(Q,t)) = 0$. This determinant is a polynomial in the entries of $m(Q,t)$. We want to find the values of t for which it vanishes.

In general, suppose $P(t)$ is some polynomial in the matrix coefficients of transport matrices $m(Q,t)$ for various points Q. Then what does the set of zeros of $P(t)$ look like, asymptotically near infinity? We will see how to recover this information from the Laplace transform of $P(t)$. From the considerations in the previous section, the Laplace transform of $P(t)$ has an analytic continuation with locally finite branching and quasi-regular singularities, and if the matrix B has been multiplied by a generic number, then the Laplace transform of $P(t)$ has faithful expansions.

So fix throughout the rest of this section the assumption that $P(t)$ is an entire function of order ≤ 1, and that its Laplace transform $f(\zeta)$ has an extension with locally finite quasi-regular singularities, with faithful expansions. Let $\mathcal{H}$ denote the convex hull of exponents of $P(t)$ (see §2).

The first task is to obtain an asymptotic expression for $P(t)$ valid on any sector. The definition of such a thing is slightly complicated by the fact that, although $P(t)$ is single valued, the terms in the asymptotic expression include

$t^{-j/N}$ and $(\log t)^k$, which may be multivalued. Let us first write down the formal asymptotic series:

$$\hat{P}(t) = \sum_{i,j,k} c_{ijk} e^{\lambda_i t} t^{-j/N} (\log t)^k.$$

The exponents λ_i are on the boundary of the hull of exponents $\mathcal{H}$. Roughly speaking, this asymptotic series should be interpreted in the following way. Any given term is negligible on some sector, the growth of $e^{\lambda_i t}$ being too small in that direction. Outside of this sector, we may choose branches of $t^{1/N}$ and $\log t$. The constants c_{ijk} will depend on these choices of branches. As in Proposition 2.3, we obtain a series $\hat{P}(t)$ which is an asymptotic series for $P(t)$, uniformly on any sector.

We may formalize this by saying that an *asymptotic series* $\hat{P}$ consists of a covering $\{U_\alpha\}$ of $\mathbf{C} - \{0\}$ by open sectors, and series $\hat{P}_\alpha(t)$ well defined on the sectors U_α respectively, such that $\hat{P}_\alpha(t)$ and $\hat{P}_\beta(t)$ are asymptotically equal on $\overline{U}_{\alpha\beta}$ (measuring the asymptotic behaviour with respect to the exponential behaviour of the leading terms in the series).

Proposition 12.1 *Suppose $P(t)$ is an entire function whose Laplace transform $f(\zeta)$ has an extension with locally finite branching, quasi-regular singularities, and faithful expansions. Then there is an asymptotic series $\hat{P}$ such that $\hat{P}_\alpha(t)$ is an asymptotic series for $P(t)$ on $\overline{U}_\alpha$.*

Proof: The same as the proof of Proposition 2.3.

We need to discuss the leading terms in the series and the overall growth of the series. For each i, let j_i be the smallest j with $c_{ijk} \neq 0$ for some k, and then let k_i be the largest such k. Thus the leading term of the series with exponent λ_i is

$$L_i(t) = c_{ij_ik_i} e^{\lambda_i t} t^{-j_i/N} (\log t)^{k_i}.$$

Set

$$W_i(t) = \left| e^{\lambda_i t} t^{-j_i/N} \right| (\log |t|)^{k_i}.$$

It is a single valued function, positive for $|t| > 1$. Thus $|L_i(t)| \leq C W_i(t)$ for $|t| \geq C$. Now, say that λ_i is a *leading exponent* if, for some $\epsilon > 0$, $W_i(t)$ is not bounded by $\epsilon \sum_{i' \neq i} W_{i'}(t)$ at large $|t|$. Say that λ_i is a *corner exponent* if, for every $C > 0$, $W_i(t)$ is not bounded by $C \sum_{i' \neq i} W_{i'}(t)$. The set of corner exponents is the minimal set of exponents such that

$$P(t) \leq C \sum_{\text{corner } i} W_i(t).$$

Let $W(t) = \sum_{corner\, i} W_i(t)$. The set of leading exponents is the minimal set of exponents such that

$$\frac{|P(t) - \sum_{leading\, i} L_i(t)|}{W(t)} \to 0$$

as $|t \to \infty|$. Let

$$L(t) = \sum_{leading\, i} L_i(t).$$

As described above, L should really be taken as a collection of series L_α well defined on sectors U_α and agreeing asymptotically on the overlaps.

The statement that $\hat{P}$ is an asymptotic series for $P(t)$ means that for any ℓ there are partial sums $\hat{P}_\alpha^\ell$ and constants C_1 and C_2 such that

$$|P(t) - \hat{P}_\alpha^\ell(t)| \le C_1 |t|^{-\ell} W(t)$$

for $t \in U_\alpha$, $|t| \ge C_2$. The fact that $L(t)$ is the leading term means that there are constants C_1, C_2, and ε such that

$$|P(t) - L_\alpha(t)| \le C_1 (\log|t|)^{-\varepsilon} W(t)$$

for $t \in U_\alpha$, $|t| \ge C_2$.

The corner exponents are arranged around the boundary of $\mathcal{H}$. Any corner of the boundary of $\mathcal{H}$ is a corner exponent. There may be corner exponents in the middle of segments on the boundary of $\mathcal{H}$, but there may also be exponents on the boundary of $\mathcal{H}$ which are not corner or even leading exponents. This depends on the powers of t and $\log t$ which occur at the particular exponent.

Denote by $\{\lambda_h\}_{h=1,\dots,r}$ the set of corner exponents on the boundary of $\mathcal{H}$, arranged in counterclockwise order. Let e_h denote the line segment between λ_{h-1} and λ_h (make the obvious convention that $h = 0$ is the same as $h = r$). We will make some constructions which are properly thought of as pertaining to the segments e_h rather than the corners h; this is the rationale for the evidently spurious notation e_h. For each segment e_h, define a curve in the complex plane

$$w(e_h) = \{t \in \mathbf{C} : W_{h-1}(t) = W_h(t) \text{ and } \Im t(\lambda_h - \lambda_{h-1}) \ge C\}.$$

Set $\lambda'(e_h) = \lambda_h - \lambda_{h-1}$ and similarly define $j'(e_h)$ and $k'(e_h)$. Then the curve $w(e_h)$ is defined by

$$\Re(t\lambda'(e_h)) = \log|t| j'(e_h)/N - \log\log|t| k'(e_h).$$

This may be graphed using $|t|$ and $\Re(t\lambda'(e_h))$ as coordinates.

The following lemma identifies the locations of the zeros of $P(t)$. They fall into series of zeros lying in bands near the curves $w(e_h)$.

Lemma 12.2 *There is a constant K such that any zero of $P(t)$ lies within the band $Z(e_h, K) = \{t : d(t, w(e_h)) \leq K\}$ for some edge e_h.*

Proof: Let $V(h, K)$ denote the region between $Z(e_h, K)$ and $Z(e_{h+1}, K)$. In the region $V(h, K)$, the corner term $L_h(t)$ is dominant. Hence, for any constant A, we may choose K large enough so that $W_h(t) > A\sum_{i\neq h} W_i(t)$ for $t \in V(h, K)$. Fix a sector U_α, and let $R_\alpha(t) = P(t) - L_\alpha(t)$ be the remainder term. On the sector U_α, there is a constant c with $|L_{\alpha,i}(t)| \geq cW_i(t)$. Hence by arranging the constant K we may assume that there is a constant a with $|L_\alpha(t)| \geq aW(t)$ for $t \in V(h, K) \cap U_\alpha$. By the estimate $|R_\alpha(t)| \leq C(\log|t|)^{-\epsilon}W(t)$, we may assume that $|L_\alpha(t)| > |R_\alpha(t)|$, so $P(t) \neq 0$ on $V(h, K)$. This works for all h, so the zeros of $P(t)$ are contained in the union of bands $Z(e_h, K)$.

We can now use the higher terms in the asymptotic series to get higher order asymptotic information on the location of the zeros of $P(t)$.

Lemma 12.3 *Suppose $f(z)$ is a holomorphic function defined in the unit disc $|z| < 1$, uniformly bounded by a constant $|f(z)| \leq C$. Suppose that for each z there is a number $n(z)$ with*

$$\left|\frac{d^{n(z)}}{dz^{n(z)}} f(z)\right| \geq n(z)!,$$

such that all $n(z)$ are less than some number N. There are constants C_1 and ϵ such that if $|f(0)| \leq \epsilon$ then there is a zero $f(z) = 0$ with $|z| \leq C_1\epsilon^{1/N}$.

Proof: We apply Newton's method, using the appropriate large derivative at each point. For each z, let $F_z(x)$ denote the sum of the first $n(z) + 1$ terms in the Taylor expansion for f at z,

$$F_z(x) = \sum_{k=0}^{n(z)} \frac{(x-z)^k}{k!} \frac{d^k f}{dz^k}(z).$$

Let $R_z(x)$ denote the remainder

$$R_z(x) = \int_z^x \frac{(y-z)^{n(z)}}{n(z)!} \frac{d^{n(z)+1} f(y)}{dy^{n(z)+1}} dy.$$

Now under the hypotheses of the lemma, define a sequence of points $z_0, z_1, \ldots$ inductively, starting with

$$z_0 = 0.$$

Having chosen z_i, let z_{i+1} denote a solution of $F_{z_i}(z_{i+1}) = 0$ nearest to z_i. If there are several possibilities, choose one. Let $\epsilon_i = |f(z_i)|$.

Now F_{z_i} is a polynomial of degree $n(z_i) \leq N$, whose highest coefficient has absolute value at least 1, and whose value at z_i is less than ϵ_i. The product of all of the roots is less than ϵ_i, so at least one root z_{i+1} is small:

$$|z_{i+1} - z_i| \leq C_0(\epsilon_i)^{1/n(z_i)}.$$

Note that $\epsilon_{i+1} = |R_{z_i}(z_{i+1})|$. The estimate on the absolute value of f gives, by Cauchy's formula, an estimate for the derivatives (we may go to a slightly smaller disc). From the above estimate and the remainder formula,

$$\epsilon_{i+1} \leq C_1(\epsilon_i)^{\frac{n(z_i)+1}{n(z_i)}}.$$

Since $n(z_i) \leq N$, we get

$$\epsilon_{i+1} \leq C_1(\epsilon_i)^{1+\frac{1}{N}},$$

so $\epsilon_i \leq C_2\epsilon^{i/N}$ (in fact the estimate is much better than this). Therefore the sequence of points z_i is Cauchy. The limit is a point z with $f(z) = 0$ and $|z| \leq C_3\epsilon^{1/N}$ as desired.

Lemma 12.4 *There are a number m_0 and constants c and ϵ such that for any point t, there is an m with $0 \leq m < m_0$ such that the absolute value of the mth derivative of the leading term is large, $|L^{(m)}(t)| \geq cW(t)$.*

Proof: We have

$$L^{(m)}(t) = \sum_i \lambda_i^m c_{ij_ik_i} e^{\lambda_i t} t^{-j_i/N} (\log t)^{k_i} + r_m(t),$$

with $|r_m(t)| \leq C(\log|t|)^{-\epsilon} W(t)$. Let m_0 be the number of terms in the sum. Since there is only one term for each distinct exponent λ_i, the Vandermonde determinant does not vanish and we may find an inverse (ν_{im}) of the matrix (λ_i^m). Thus

$$c_{ij_ik_i} e^{\lambda_i t} t^{-j_i/N} (\log t)^{k_i} = \sum_m \nu_{im}(L^{(m)}(t) - r_m(t)).$$

Since the vector on the left has size comparable to $W(t)$, there is a constant c_1 such that $|L^{(m)}(t) - r_m(t)| \geq c_1 W(t)$ for some m. By the estimate for $r_m(t)$, there is a smaller constant c and a large C_2 such that if $|t| \geq C_2$, then $|L^{(m)}(t)| \geq cW(t)$. Finally, increase m_0 and decrease c to treat the region $|t| \leq C_2$

Corollary 12.5 *There are a number m_0 and constants c and ϵ such that for any point t, there is an m with $0 \leq m < m_0$ such that $|P^{(m)}(t)| \geq cW(t)$. The same is true for any partial sum of the expansion $\hat{P}$.*

Proof: Given ϵ, we may choose C so that for $|t| \geq C$, the remainder is small $|R(t)| \leq \epsilon W(t)$. If $|t - t_0| \leq 1$ then $W(t) \leq cW(t_0)$. By Cauchy's formula we get the same estimate for the derivatives, $|R^{(m)}(t)| \leq \epsilon W(t)$ for $|t| \geq C$. Again, the region $|t| \leq C$ can be treated separately. Combining this with the lemma gives the corollary.

Given a number ℓ, let $\hat{P}_h^\ell$ denote the partial sum of the series involving only exponents on the full straight line segment in the boundary of $\mathcal{H}$ containing λ_{h-1} and λ_h. This is a valid approximation on the band $Z(e_h, K)$. Let $w^\ell(e_h)$ be the set of zeros of $\hat{P}_h^\ell$ in the band $Z(e_h, K)$.

Let m_0 be the number given by Lemma 12.4. Consider the union of small discs

$$Z^\ell(e_h, K) = \bigcup_{t_0 \in w^\ell(e_h)} D(t_0, K|t|^{-\ell/m_0}).$$

The following proposition gives an asymptotic description of the location of the zeros of $P(t)$.

Proposition 12.6 *For any ℓ, we can choose K such that the following holds. If t is a zero of $P(t)$ in the band $Z(e_h, K)$, then t is contained in $Z^\ell(e_h, K)$. Furthermore, in a given connected component of $Z^\ell(e_h, K)$ the number of zeros of $P(t)$ (counted with multiplicity) is the same as the number of zeros of $\hat{P}_h^\ell(t)$.*

Proof: Write $P(t) = \hat{P}_h^\ell(t) + R(t)$ with $|R(t)| \leq C|t|^{-\ell}$. If t_0 is a point in $Z(e_h, K)$ with $|P(t_0)| \leq C|t_0|^{-\ell}$, then $|\hat{P}_h^\ell(t_0)| \leq C|t_0|^{-\ell}$, so by applying Lemma 12.3 (rescaling everything by dividing by $W(t_0)$) we get a point t_1 with $\hat{P}_h^\ell(t_1) = 0$ (so $t_1 \in w^\ell(e_h)$) and $|t_0 - t_1| \leq C|t_0|^{-\ell/m_0}$. Thus t_0 is in $Z^\ell(e_h, K_1)$. Applying this to zeros t_0 of $P(t)$ we get the first statement. Applying it to points on the boundary of a connected component of $Z^\ell(e_h, K_1)$, we see that K_1 may be chosen so that $|R(t)| < |P(t)|$ on the boundary. Hence the winding numbers of $P(t)$ and $\hat{P}_h^\ell(t)$ are the same, which gives the second part.

Remark: One can make a similar statement saying that the zeros of $P(t)$ are near zeros of $L(t)$, to order $(\log|t|)^{-\varepsilon}$ for some $\varepsilon > 0$.

Proposition 12.7 *The zeros of $P(t)$ are approximately evenly distributed in the band $Z(e_h, K)$, with density equal to $(2\pi)^{-1}$ times the length of the segment*

$|e_h| = |\lambda_h - \lambda_{h-1}|$. *In other words, there is a constant C such that the number $N(e_h, R)$ of zeros in $Z(e_h, K)$ at distance $\leq R$ from the origin satisfies*

$$|e_h|R - C \leq 2\pi N(e_h, R) \leq |e_h|R + C.$$

Proof: First treat $L(t)$. We may assume $h = 1$ so λ_0 and λ_1 are the exponents on the clockwise and counterclockwise ends of the segment e_1. By making a rotation in the t plane, we can assume that $\lambda' = \lambda_1 - \lambda_0$ is purely imaginary. The curve $w(e_h)$ goes to infinity in approximately the positive real direction (there may be logarithmic and doubly logarithmic corrections).

The term $e^{\lambda_1 t} t^{-j_1/N} (\log t)^{k_1}$ is dominant on the positive imaginary side of the band $Z(e_1, K)$. It has argument approximately $\Im(\lambda_1 t)$, since the argument of t is approximately zero and the argument of $\log t$ is always almost zero. On the negative imaginary side of the band $Z(e_1, K)$ the term $e^{\lambda_0 t} t^{-j_0/N} (\log t)^{k_0}$ is dominant, with argument approximately $\Im(\lambda_0 t)$. These approximations are correct to within an error which goes to zero as $|t| \to \infty$. The error due to the other terms in $L(t)$ is bounded as $|t| \to \infty$, and can be made small by taking the width of the band to be small. The difference between the arguments on the two sides of the band is $\Im(\lambda_1 - \lambda_0)t$. Since the real parts of λ_0 and λ_1 coincide, there is no contribution from the possibly logarithmic imaginary part of t. Since the imaginary part of t is logarithmic in the real part of t, $(|t| - \Re t) \to 0$. The difference between the arguments is therefore equal to $\Im(\lambda_1 - \lambda_0)|t|$.

The difference between the winding numbers on the two sides of the band is equal to the number of zeros within the band. We have to show that in moving from one side of the band to the other, the argument changes by at most a bounded amount. Let $\lambda^1, \ldots, \lambda^m$ be the exponents occuring in the part of the leading term $L(t)$ which is dominant near $w(e_1)$. Their real parts are all equal; they lie in the line segment between λ_0 and λ_1 (which two are also included somewhere in the list). Then near $w(e_1)$, $L(t)$ is equal to a sum of terms with arguments $\Im(\lambda^1 t), \ldots, \Im(\lambda^m t)$. Fix a path across the band $Z(e_1, K)$, given by $\Re t = u$. The imaginary part $v = \Im t$ moves from a to $a + K$, where a is some number logarithmic in u. The arguments of the terms are approximately $u\Im\lambda^i + v\Re\lambda^i$ (and the error goes to zero as $|t| \to \infty$). The $v\Re\lambda^i$ are all the same. We contend that there is a constant C such that for any value u_0, there is a value u between u_0 and $u_0 + C$ such that the $u\Im\lambda^i$ are all contained in the same quarter-circle (i.e. interval of length $< \pi/2$). Consider the points $u(\Im\lambda^1, \ldots, \Im\lambda^m)$ in the torus $(\mathbf{R}/2\pi\mathbf{Z})^m$. The set U of points $(\theta_1, \ldots, \theta_m)$ where $\theta_1, \ldots, \theta_m$ are all contained in the same quarter-circle, is an open neighborhood of the origin. Let T be the torus which is the closure of the set of points $u(\Im\lambda^1, \ldots, \Im\lambda^m)$. There is a constant C such that T can be covered the union of translates $U + w(\Im\lambda^1, \ldots, \Im\lambda^m)$ for $w \in [0, C]$. So, given any point u_0, there is a u such that $u - u_0 \in [0, C]$ and $(u\Im\lambda^1, \ldots, u\Im\lambda^m) \in U$, as contended. Therefore, along the segment $\Re t = u$, the arguments of the terms in $L(t)$ are all contained in the same quarter-circle. Hence there is no cancellation, and the argument of the sum $L(t)$ is contained in that quarter-circle. So for these paths across the band $Z(e_1, K)$, the argument changes by no more than π. Since these good paths occur regularly (at least one in every interval of length $\leq C$), the estimate on the winding number obtained by considering the winding numbers on the positive and negative imaginary sides of the band is correct up to a bounded error.

This proves the proposition for $L(t)$ (taking into account the fact that winding number equals argument divided by 2π). The proposition for $P(t)$ follows immediately, because it differs from $L(t)$ by a remainder which is comparatively arbitrarily small. Note that on the good paths going across the band $Z(e_1, K)$, the arguments of all terms in $L(t)$ are contained in the same quarter-circle. Thus there is no cancellation, and the size of $L(t)$ is as big as $W(t)$. Hence the size of the remainder is small compared to the size of $L(t)$.

References

[1] R. Beals, P. Deift, C. Tomei. *Direct and Inverse Scattering on the Line. AMS Math. Surveys and Monographs* **28** (1988).

[2] G. D. Birkhoff. On the asymptotic character of the solutions of certain linear differential equations containing a parameter. *Trans. Amer. Math. Soc.* **9** (1908), 219-231.

[3] G. D. Birkhoff. Boundary value and expansion problems of ordinary linear differential equations. *Trans. Amer. Math. Soc.* **9** (1908), 373-395.

[4] K-T. Chen. Integration of paths, geometric invariants and a generalized Baker-Hausdorff formula, *Ann. of Math.* **65** (1957), 163-178.

[5] C. H. Clemens. Degeneration of Kähler manifolds.. *Duke Math. J.* **44** (1977), 215-290.

[6] W. Eckhaus. *Asymptotic Analysis of Singular Perturbations* North Holland, Amsterdam (1979).

[7] A. G. Eliseev and S. A. Lomov. Theory of singular perturbations in the case of spectral singularities of the limit operator. *Math. USSR Sbornik* **131** (1986), English translation in vol. **59**, 541-555.

[8] H. Federer. *Geometric Measure Theory.* Springer-Verlag, Berlin (1969).

[9] H. Gingold. An invariant asymptotic formula for solutions of second-order linear ODE's. *Asymptotic Analysis* **1** (1988), 317-350.

[10] J. Grasman. *Asymptotic Methods for Relaxation Oscillators and Applications. Applied Mathematical Sciences* **63** Springer, New York (1987).

[11] E. Hille. *Lectures on Differential Equations.* Addison-Wesley, Reading, Mass., (1969).

[12] H. Hironaka. Resolution of singularities of an algebraic variety over a field of characteristic zero, I and II. *Ann. of Math.* **79** (1964), 109-326.

[13] V. I. Kachalov and S. A. Lomov, Smoothness of solutions of differential equations with respect to a singular parameter. *Sov. Math. Doklady* **37** (1988), 465-467.

[14] V. I. Kachalov and S. A. Lomov, On the analytic properties of solutions of differential equations with singular points. *Sov. Math. Doklady* **39** (1989), 12-14.

[15] M. Kruskal. Asymptotic theory of Hamiltonian and other systems with all solutions nearly periodic. *J. Math. Phys.* **3** no. 4 (1962), 806ff.

[16] A. Landman. On the Picard-Lefschetz transformation for algebraic manifolds acquiring general singularities. *Trans. Amer. Math. Soc.* **18** (1973), 89-126.

[17] R. E. Langer. The asymptotic solutions of ordinary linear differential equations of the second order with special reference to a turning point. *Trans. Amer. Math. Soc.* **67** (1949), 461-490.

[18] G. Laumon. Transformation de Fourier constantes d'équations fonctionelles et conjecture de Weil. *Publ. Math. I. H. E. S.* **65** (1987), 131-210.

[19] J. Liouville. Mémoire sur le développement des fonctions ou parties de fonctions en séries dont divers terms sont assujettis à satisfaire à une même équation différentielle du second ordre contenant un paramètre variable. *J. Math. Pures et Appl.* **1** (1836), 253-266.

[20] J. Liouville. Second mémoire sur le développement des fonctions ou parties de fonctions en séries dont divers terms sont assujettis à satisfaire à une même équation différentielle du second ordre contenant un paramètre variable. *J. Math. Pures et Appl.* **2** (1837), 16-35.

[21] J. Liouville. Premier mémoire sur la théorie des équations différentielles linéaires et sur le développement des fonctions en séries. *J. Math. Pures et Appl.* **3** (1838), 561-614.

[22] Mishchenko and Rozov. *Differential Equations With Small Parameters and Relaxation Oscillations.* Plenum Press, New York (1980).

[23] A. H. Nayfeh. *Introduction to Perturbation Techniques.* Wiley, New York (1981).

[24] K. Nipp. An extension of Tikhonov's theorem in singular perturbations for the planar case. *Z. Angew. Math. Phys.* **34** (1983), 277-290.

[25] R. E. O'Malley. Topics in singular perturbations. *Adv. in Math.* **2**, Acad. Press (1968).

[26] E. Picard. Sur l'application des méthodes d'approximations successives à l'étude de certaines equations différentielles ordinaires. *J. Math. Pures et Appl.* (2 ser.) **IX**, (1893), 217-271.

[27] L. S. Pontryagin. Asymptotic behavior of the solutions of systems of differential equations with a small parameter in the higher derivatives. *A. M. S. Translations* **18** (1961), 295-319.

[28] C. Procesi. The invariant theory of $n \times n$ matrices. *Adv. in Math.* **19** (1976), 306-381.

[29] C. T. Simpson. Moduli of representations of the fundamental group of a smooth projective variety. Preprint, Princeton University (1989).

[30] C. T. Simpson. Transcendental aspects of the Riemann-Hilbert correspondence. To appear, *Illinois Math. J.* (1990).

[31] C. Sturm. Mémoire sur les équations différentielles linéaires du second ordre. *J. Math. Pures et Appl.* **1** (1836), 106-186.

[32] A. N. Tikhonov. Systems of differential equations containing small parameters in the derivatives. *Mat. Sbornik* **31** (1952), 574-584.

[33] W. R. Wasow. *Asymptotic Expansions for Ordinary Differential Equations.* Interscience, New York (1965).

[34] W. R. Wasow. *Linear Turning Point Theory.* Springer, Berlin (1984).

[35] W. R. Wasow. The capriciousness of singular perturbations. *Nieuv. Arch. Wisk.* **18** (1970), 190-210.

[36] E. T. Whittaker and G. N. Watson. *A Course of Modern Analysis.* Cambridge University Press, Cambridge, (1902, ...).

References added in proof:

[37] J. Ecalle. Les fonctions resurgentes. *Publ. Math. Université de Paris-Sud*, several volumes.

[38] E. Delabaere, H. Dillinger. Contribution à la résurgence quantique: Résurgence de Voros et Fonction spectral de Jost. Doctoral thesis in Mathematics (under the direction of F. Pham), Université de Nice Sophia-Antipolis (1991).

Index

Vol. 1411: B. Jiang (Ed.), Topological Fixed Point Theory and Applications. Proceedings. 1988. VI, 203 pages. 1989.

Vol. 1412: V.V. Kalashnikov, V.M. Zolotarev (Eds.), Stability Problems for Stochastic Models. Proceedings, 1987. X, 380 pages. 1989.

Vol. 1413: S. Wright, Uniqueness of the Injective III_1 Factor. III, 108 pages. 1989.

Vol. 1414: E. Ramirez de Arellano (Ed.), Algebraic Geometry and Complex Analysis. Proceedings, 1987. VI, 180 pages. 1989.

Vol. 1415: M. Langevin, M. Waldschmidt (Eds.), Cinquante Ans de Polynômes. Fifty Years of Polynomials. Proceedings, 1988. IX, 235 pages.1990.

Vol. 1416: C. Albert (Ed.), Géométrie Symplectique et Mécanique. Proceedings, 1988. V, 289 pages. 1990.

Vol. 1417: A.J. Sommese, A. Biancofiore, E.L. Livorni (Eds.), Algebraic Geometry. Proceedings, 1988. V, 320 pages. 1990.

Vol. 1418: M. Mimura (Ed.), Homotopy Theory and Related Topics. Proceedings, 1988. V, 241 pages. 1990.

Vol. 1419: P.S. Bullen, P.Y. Lee, J.L. Mawhin, P. Muldowney, W.F. Pfeffer (Eds.), New Integrals. Proceedings, 1988. V, 202 pages. 1990.

Vol. 1420: M. Galbiati, A. Tognoli (Eds.), Real Analytic Geometry. Proceedings, 1988. IV, 366 pages. 1990.

Vol. 1421: H.A. Biagioni, A Nonlinear Theory of Generalized Functions, XII, 214 pages. 1990.

Vol. 1422: V. Villani (Ed.), Complex Geometry and Analysis. Proceedings, 1988. V, 109 pages. 1990.

Vol. 1423: S.O. Kochman, Stable Homotopy Groups of Spheres: A Computer-Assisted Approach. VIII, 330 pages. 1990.

Vol. 1424: F.E. Burstall, J.H. Rawnsley, Twistor Theory for Riemannian Symmetric Spaces. III, 112 pages. 1990.

Vol. 1425: R.A. Piccinini (Ed.), Groups of Self-Equivalences and Related Topics. Proceedings, 1988. V, 214 pages. 1990.

Vol. 1426: J. Azéma, P.A. Meyer, M. Yor (Eds.), Séminaire de Probabilités XXIV, 1988/89. V, 490 pages. 1990.

Vol. 1427: A. Ancona, D. Geman, N. Ikeda, École d'Eté de Probabilités de Saint Flour XVIII, 1988. Ed.: P.L. Hennequin. VII, 330 pages. 1990.

Vol. 1428: K. Erdmann, Blocks of Tame Representation Type and Related Algebras. XV. 312 pages. 1990.

Vol. 1429: S. Homer, A. Nerode, R.A. Platek, G.E. Sacks, A. Scedrov, Logic and Computer Science. Seminar, 1988. Editor: P. Odifreddi. V, 162 pages. 1990.

Vol. 1430: W. Bruns, A. Simis (Eds.), Commutative Algebra. Proceedings. 1988. V, 160 pages. 1990.

Vol. 1431: J.G. Heywood, K. Masuda, R. Rautmann, V.A. Solonnikov (Eds.), The Navier-Stokes Equations – Theory and Numerical Methods. Proceedings, 1988. VII, 238 pages. 1990.

Vol. 1432: K. Ambos-Spies, G.H. Müller, G.E. Sacks (Eds.), Recursion Theory Week. Proceedings, 1989. VI, 393 pages. 1990.

Vol. 1433: S. Lang, W. Cherry, Topics in Nevanlinna Theory. II, 174 pages.1990.

Vol. 1434: K. Nagasaka, E. Fouvry (Eds.), Analytic Number Theory. Proceedings, 1988. VI, 218 pages. 1990.

Vol. 1435: St. Ruscheweyh, E.B. Saff, L.C. Salinas, R.S. Varga (Eds.), Computational Methods and Function Theory. Proceedings, 1989. VI, 211 pages. 1990.

Vol. 1436: S. Xambó-Descamps (Ed.), Enumerative Geometry. Proceedings, 1987. V, 303 pages. 1990.

Vol. 1437: H. Inassaridze (Ed.), K-theory and Homological Algebra. Seminar, 1987–88. V, 313 pages. 1990.

Vol. 1438: P.G. Lemarié (Ed.) Les Ondelettes en 1989. Seminar. IV, 212 pages. 1990.

Vol. 1439: E. Bujalance, J.J. Etayo, J.M. Gamboa, G. Gromadzki. Automorphism Groups of Compact Bordered Klein Surfaces: A Combinatorial Approach. XIII, 201 pages. 1990.

Vol. 1440: P. Latiolais (Ed.), Topology and Combinatorial Groups Theory. Seminar, 1985–1988. VI, 207 pages. 1990.

Vol. 1441: M. Coornaert, T. Delzant, A. Papadopoulos. Géométrie et théorie des groupes. X, 165 pages. 1990.

Vol. 1442: L. Accardi, M. von Waldenfels (Eds.), Quantum Probability and Applications V. Proceedings, 1988. VI, 413 pages. 1990.

Vol. 1443: K.H. Dovermann, R. Schultz, Equivariant Surgery Theories and Their Periodicity Properties. VI, 227 pages. 1990.

Vol. 1444: H. Korezlioglu, A.S. Ustunel (Eds.), Stochastic Analysis and Related Topics VI. Proceedings, 1988. V, 268 pages. 1990.

Vol. 1445: F. Schulz, Regularity Theory for Quasilinear Elliptic Systems and – Monge Ampère Equations in Two Dimensions. XV, 123 pages. 1990.

Vol. 1446: Methods of Nonconvex Analysis. Seminar, 1989. Editor: A. Cellina. V, 206 pages. 1990.

Vol. 1447: J.-G. Labesse, J. Schwermer (Eds), Cohomology of Arithmetic Groups and Automorphic Forms. Proceedings, 1989. V, 358 pages. 1990.

Vol. 1448: S.K. Jain, S.R. López-Permouth (Eds.), Non-Commutative Ring Theory. Proceedings, 1989. V, 166 pages. 1990.

Vol. 1449: W. Odyniec, G. Lewicki, Minimal Projections in Banach Spaces. VIII, 168 pages. 1990.

Vol. 1450: H. Fujita, T. Ikebe, S.T. Kuroda (Eds.), Functional-Analytic Methods for Partial Differential Equations. Proceedings, 1989. VII, 252 pages. 1990.

Vol. 1451: L. Alvarez-Gaumé, E. Arbarello, C. De Concini, N.J. Hitchin, Global Geometry and Mathematical Physics. Montecatini Terme 1988. Seminar. Editors: M. Francaviglia, F. Gherardelli. IX, 197 pages. 1990.

Vol. 1452: E. Hlawka, R.F. Tichy (Eds.), Number-Theoretic Analysis. Seminar, 1988–89. V, 220 pages. 1990.

Vol. 1453: Yu.G. Borisovich, Yu.E. Gliklikh (Eds.), Global Analysis – Studies and Applications IV. V, 320 pages. 1990.

Vol. 1454: F. Baldassari, S. Bosch, B. Dwork (Eds.), p-adic Analysis. Proceedings, 1989. V, 382 pages. 1990.

Vol. 1455: J.-P. Françoise, R. Roussarie (Eds.), Bifurcations of Planar Vector Fields. Proceedings, 1989. VI, 396 pages. 1990.

Vol. 1456: L.G. Kovács (Ed.), Groups – Canberra 1989. Proceedings. XII, 198 pages. 1990.

Vol. 1457: O. Axelsson, L.Yu. Kolotilina (Eds.), Preconditioned Conjugate Gradient Methods. Proceedings, 1989. V, 196 pages. 1990.

Vol. 1458: R. Schaaf, Global Solution Branches of Two Point Boundary Value Problems. XIX, 141 pages. 1990.

Vol. 1459: D. Tiba, Optimal Control of Nonsmooth Distributed Parameter Systems. VII, 159 pages. 1990.

Vol. 1460: G. Toscani, V. Boffi, S. Rionero (Eds.), Mathematical Aspects of Fluid Plasma Dynamics. Proceedings, 1988. V, 221 pages. 1991.

Vol. 1461: R. Gorenflo, S. Vessella, Abel Integral Equations. VII, 215 pages. 1991.

Vol. 1462: D. Mond, J. Montaldi (Eds.), Singularity Theory and its Applications. Warwick 1989, Part I. VIII, 405 pages. 1991.

Vol. 1463: R. Roberts, I. Stewart (Eds.), Singularity Theory and its Applications. Warwick 1989, Part II. VIII, 322 pages. 1991.

Vol. 1464: D. L. Burkholder, E. Pardoux, A. Sznitman, Ecole d'Eté de Probabilités de Saint- Flour XIX-1989. Editor: P. L. Hennequin. VI, 256 pages. 1991.

Vol. 1465: G. David, Wavelets and Singular Integrals on Curves and Surfaces. X, 107 pages. 1991.

Vol. 1466: W. Banaszczyk, Additive Subgroups of Topological Vector Spaces. VII, 178 pages. 1991.

Vol. 1467: W. M. Schmidt, Diophantine Approximations and Diophantine Equations. VIII, 217 pages. 1991.

Vol. 1468: J. Noguchi, T. Ohsawa (Eds.), Prospects in Complex Geometry. Proceedings, 1989. VII, 421 pages. 1991.

Vol. 1469: J. Lindenstrauss, V. D. Milman (Eds.), Geometric Aspects of Functional Analysis. Seminar 1989-90. XI, 191 pages. 1991.

Vol. 1470: E. Odell, H. Rosenthal (Eds.), Functional Analysis. Proceedings, 1987-89. VII, 199 pages. 1991.

Vol. 1471: A. A. Panchishkin, Non-Archimedean L-Functions of Siegel and Hilbert Modular Forms. VII, 157 pages. 1991.

Vol. 1472: T. T. Nielsen, Bose Algebras: The Complex and Real Wave Representations. V, 132 pages. 1991.

Vol. 1473: Y. Hino, S. Murakami, T. Naito, Functional Differential Equations with Infinite Delay. X, 317 pages. 1991.

Vol. 1474: S. Jackowski, B. Oliver, K. Pawałowski (Eds.), Algebraic Topology, Poznań 1989. Proceedings. VIII, 397 pages. 1991.

Vol. 1475: S. Busenberg, M. Martelli (Eds.), Delay Differential Equations and Dynamical Systems. Proceedings, 1990. VIII, 249 pages. 1991.

Vol. 1476: M. Bekkali, Topics in Set Theory. VII, 120 pages. 1991.

Vol. 1477: R. Jajte, Strong Limit Theorems in Noncommutative L_2-Spaces. X, 113 pages. 1991.

Vol. 1478: M.-P. Malliavin (Ed.), Topics in Invariant Theory. Seminar 1989-1990. VI, 272 pages. 1991.

Vol. 1479: S. Bloch, I. Dolgachev, W. Fulton (Eds.), Algebraic Geometry. Proceedings, 1989. VII, 300 pages. 1991.

Vol. 1480: F. Dumortier, R. Roussarie, J. Sotomayor, H. Żoładek, Bifurcations of Planar Vector Fields: Nilpotent Singularities and Abelian Integrals. VIII, 226 pages. 1991.

Vol. 1481: D. Ferus, U. Pinkall, U. Simon, B. Wegner (Eds.), Global Differential Geometry and Global Analysis. Proceedings, 1991. VIII, 283 pages. 1991.

Vol. 1482: J. Chabrowski, The Dirichlet Problem with L^2-Boundary Data for Elliptic Linear Equations. VI, 173 pages. 1991.

Vol. 1483: E. Reithmeier, Periodic Solutions of Nonlinear Dynamical Systems. VI, 171 pages. 1991.

Vol. 1484: H. Delfs, Homology of Locally Semialgebraic Spaces. IX, 136 pages. 1991.

Vol. 1485: J. Azéma, P. A. Meyer, M. Yor (Eds.), Séminaire de Probabilités XXV. VIII, 440 pages. 1991.

Vol. 1486: L. Arnold, H. Crauel, J.-P. Eckmann (Eds.), Lyapunov Exponents. Proceedings, 1990. VIII, 365 pages. 1991.

Vol. 1487: E. Freitag, Singular Modular Forms and Theta Relations. VI, 172 pages. 1991.

Vol. 1488: A. Carboni, M. C. Pedicchio, G. Rosolini (Eds.), Category Theory. Proceedings, 1990. VII, 494 pages. 1991.

Vol. 1490: K. Metsch, Linear Spaces with Few Lines. XIII, 196 pages. 1991.

Vol. 1492: K. R. Wicks, Fractals and Hyperspaces. VIII, 168 pages. 1991.

Vol. 1493: E. Benoît (Ed.), Dynamic Bifurcations. Proceedings, Luminy 1990. VII, 219 pages. 1991.

Vol. 1494: M.-T. Cheng, X.-W. Zhou, D.-G. Deng (Eds.), Harmonic Analysis. Proceedings, 1988. IX, 226 pages. 1991.

Vol. 1495: J. M. Bony, G. Grubb, L. Hörmander, H. Komatsu, J. Sjöstrand, Microlocal Analysis and Applications. Montecatini Terme, 1989. Editors: L. Cattabriga, L. Rodino. VII, 349 pages. 1991.

Vol. 1496: C. Foias, B. Francis, J. W. Helton, H. Wakernaak, J. B. Pearson, H_∞-Control Theory. Como, 1990. Editors: E. Mosca, L. Pandolfi. VII, 336 pages. 1991.

Vol. 1497: G. T. Herman, A. K. Louis, F. Natterer (Eds.), Mathematical Methods in Tomography. Proceedings 1990. X, 268 pages. 1991.

Vol. 1498: R. Lang, Spectral Theory of Random Schrödinger Operators. X, 125 pages. 1991.

Vol. 1499: K. Taira, Elliptic Boundary Value Problems. IX, 132 pages. 1991.

Vol. 1501: A. De Masi, E. Presutti, Mathematical Methods for Hydrodynamic Limits. IX, 196 pages. 1991.

Vol. 1502: C. Simpson, Asymptotic Behavior of Monodromy. V, 139 pages. 1991.